Imagine Lagos

NEW AFRICAN HISTORIES

SERIES EDITORS: JEAN ALLMAN, ALLEN ISAACMAN, DEREK R. PETERSON, AND CARINA RAY

David William Cohen and E. S. Atieno Odhiambo, *The Risks of Knowledge*
Belinda Bozzoli, *Theatres of Struggle and the End of Apartheid*
Gary Kynoch, *We Are Fighting the World*
Stephanie Newell, *The Forger's Tale*
Jacob A. Tropp, *Natures of Colonial Change*
Jan Bender Shetler, *Imagining Serengeti*
Cheikh Anta Babou, *Fighting the Greater Jihad*
Marc Epprecht, *Heterosexual Africa?*
Marissa J. Moorman, *Intonations*
Karen E. Flint, *Healing Traditions*
Derek R. Peterson and Giacomo Macola, editors, *Recasting the Past*
Moses E. Ochonu, *Colonial Meltdown*
Emily S. Burrill, Richard L. Roberts, and Elizabeth Thornberry, editors, *Domestic Violence and the Law in Colonial and Postcolonial Africa*
Daniel R. Magaziner, *The Law and the Prophets*
Emily Lynn Osborn, *Our New Husbands Are Here*
Robert Trent Vinson, *The Americans Are Coming!*
James R. Brennan, *Taifa*
Benjamin N. Lawrance and Richard L. Roberts, editors, *Trafficking in Slavery's Wake*
David M. Gordon, *Invisible Agents*
Allen F. Isaacman and Barbara S. Isaacman, *Dams, Displacement, and the Delusion of Development*
Stephanie Newell, *The Power to Name*
Gibril R. Cole, *The Krio of West Africa*
Matthew M. Heaton, *Black Skin, White Coats*
Meredith Terretta, *Nation of Outlaws, State of Violence*
Paolo Israel, *In Step with the Times*
Michelle R. Moyd, *Violent Intermediaries*
Abosede A. George, *Making Modern Girls*
Alicia C. Decker, *In Idi Amin's Shadow*
Rachel Jean-Baptiste, *Conjugal Rights*
Shobana Shankar, *Who Shall Enter Paradise?*
Emily S. Burrill, *States of Marriage*
Todd Cleveland, *Diamonds in the Rough*
Carina E. Ray, *Crossing the Color Line*
Sarah Van Beurden, *Authentically African*
Giacomo Macola, *The Gun in Central Africa*
Lynn Schler, *Nation on Board*
Julie MacArthur, *Cartography and the Political Imagination*
Abou B. Bamba, *African Miracle, African Mirage*
Daniel Magaziner, *The Art of Life in South Africa*
Paul Ocobock, *An Uncertain Age*
Keren Weitzberg, *We Do Not Have Borders*
Nuno Domingos, *Football and Colonialism*
Jeffrey S. Ahlman, *Living with Nkrumahism*
Bianca Murillo, *Market Encounters*
Laura Fair, *Reel Pleasures*
Thomas F. McDow, *Buying Time*
Jon Soske, *Internal Frontiers*
Elizabeth W. Giorgis, *Modernist Art in Ethiopia*
Matthew V. Bender, *Water Brings No Harm*
David Morton, *Age of Concrete*
Marissa J. Moorman, *Powerful Frequencies*
Ndubueze L. Mbah, *Emergent Masculinities*
Judith A. Byfield, *The Great Upheaval*
Patricia Hayes and Gary Minkley, editors, *Ambivalent*
Mari K. Webel, *The Politics of Disease Control*
Kara Moskowitz, *Seeing Like a Citizen*
Jacob Dlamini, *Safari Nation*
Alice Wiemers, *Village Work*
Cheikh Anta Babou, *The Muridiyya on the Move*
Laura Ann Twagira, *Embodied Engineering*
Marissa Mika, *Africanizing Oncology*
Holly Hanson, *To Speak and Be Heard*
Paul S. Landau, *Spear*
Saheed Aderinto, *Animality and Colonial Subjecthood in Africa*
Katherine Bruce-Lockhart, *Carceral Afterlives*
Natasha Erlank, *Convening Black Intimacy in Early Twentieth-Century South Africa*
Morgan J. Robinson, *A Language for the World*
Faeeza Ballim, *Apartheid's Leviathan*
Nicole Eggers, *Unruly Ideas*
Mark W. Deets, *A Country of Defiance*
Patrick William Otim, *Acholi Intellectuals*
Ademide Adelusi-Adeluyi, *Imagine Lagos*

Imagine Lagos

Mapping History, Place, and Politics in a Nineteenth-Century African City

Ademide Adelusi-Adeluyi

OHIO UNIVERSITY PRESS

ATHENS, OHIO

Ohio University Press, Athens, Ohio 45701
ohioswallow.com

Printed in the United States of America
Ohio University Press books are printed on acid-free paper ∞ ™

Library of Congress Cataloging-in-Publication Data
Names: Adelusi-Adeluyi, Ademide, author.
Title: Imagine Lagos : mapping history, place, and politics in a nineteenth-century African city / Ademide Adelusi-Adeluyi.
Other titles: Mapping history, place, and politics in a nineteenth-century African city | New African histories series.
Description: Athens, Ohio : Ohio University Press, [2024] | Series: New African histories | Includes bibliographical references and index. | Contents: Introduction—Dots and Lines on a Map: A Note on Method—Streets, Placemaking, and History in Old Lagos—Who Broke Lagos?—A New Eko?— Recovering Lost Ground in Old Lagos—Placing Justice—Conclusion: Eko o ni Ranti?
Identifiers: LCCN 2023016535 (print) | LCCN 2023016536 (ebook) | ISBN 9780821424896 (paperback) | ISBN 9780821424889 (hardback) | ISBN 9780821447802 (adobe pdf)
Subjects: LCSH: City and town life—Nigeria—Lagos—History—19th century. | Streetscapes (Urban design)—Nigeria—Lagos—History—19th century. | Lagos (Nigeria)—History—19th century.
Classification: LCC DT515.9.L3 A34 2023 (print) | LCC DT515.9.L3 (ebook) | DDC 966.910901—dc23/eng/20230411
LC record available at https://lccn.loc.gov/2023016535
LC ebook record available at https://lccn.loc.gov/2023016536

Imagine Lagos is the recipient of the
Ohio University Press First Book Fund,
established by Gillian Berchowitz in 2018.

First-time authors are often pioneers in their fields, and their writing and research are crucial to understanding the critical issues of our time. The Ohio University Press First Book Fund sets out to make the process more equitable for African and Appalachian scholars as they seek to publish their first books.

For more information, please see ohioswallow.com/first-book-fund/.

Contents

Illustrations

MAPS

FIGURES

Acknowledgments

This book has been many years in the making, and as such has acquired at least a decade of debts as it has evolved. First, many thanks to my graduate school adviser at New York University (NYU), Michael Gomez. Since I began as a master's student at NYU, Michael has been a source of inspiration, encouragement, and support. He saw my potential before I understood what writing history really meant. He encouraged me to pursue history professionally, and I will always be grateful for his unwavering support. The members of my dissertation committee, Fred Cooper, Abosede George, Saheed Aderinto, and Guy Ortolano, were inspiring and excellent, and provided me with ideas and a direction to transform my dissertation into this book.

I received so much support from faculty, staff, and my peers in the graduate program at NYU. I offer my thanks to Thomas Bender, Karin Burell, Robinette Dowtin, Rashauna Johnson, Ebony Jones, Larissa Kopytoff, Evelyne Laurent-Perrault, Tyesha Maddox, Yuko Miki, Alaina Morgan, Alison Okuda, Nathalie Pierre, Daniel Rodriguez, and Barbara Weinstein. My colleagues at the NYU Center for Humanities (then the Humanities Initiative), where I spent a very constructive year revising my dissertation, were instrumental in helping me shift my research emphasis from an urban history to one that was more grounded in spatial analysis. Dwai Banerjee, J. de Leon, Dania Hueckmann, Gwyneth Malin, Andy Romig, Cara Shousterman, Delia Solomons, Zeb Tortorici, and Jane Tylus were tireless in their feedback on various early ideas and versions of the project.

I spent the 2016–17 academic year at the Humanities Research Center at Rice University, with the generous support of a postdoctoral fellowship award from the Andrew W. Mellon Foundation. I began the new maps for this project there, and I am thankful for the support from Alex Byrd, Farès El-Dahdah, Alida Metcalf, and Jean Aroom. My cohort and peers there—Carroll Parrott Blue, Shannon Iverson, Aime von Bokel, Guan Pei, and Rex Troumbley—were also an important part of my time in Houston.

I have workshopped these chapters with several groups, and I am especially thankful to those who read the manuscript in the workshop funded by the University of California Humanities Research Institute. The workshop happened just as the COVID-19 pandemic suddenly shut down California and much of the world in March 2020. I remain grateful to all the readers who were so flexible as I quickly transformed a two-day in-person workshop to a five-hour virtual one. These readers, including Ray Kea, S. Wright Kennedy, Laura Mitchell, Ángel David Nieves, Moses Ochonu, Harry Odamtten, and Lorelle Semley, offered generative and insightful feedback on the draft of the book. I also offer my thanks to Jamie Warren, anonymous readers at the press for their insightful feedback on the penultimate draft of the book, and to Jean Allman, Carina Ray, Tyler Balli, Beth Pratt, and the team at Ohio University Press for their support to this project.

Funding from the Hellman Family Foundation and University of California Regents' Award for Junior Faculty supported my research in Nigeria and in the UK. A Career Enhancement Fellowship for Junior Faculty from the Citizens and Scholars Foundation (formerly the Woodrow Wilson National Fellowship Foundation) supported my research and provided me valuable time off to write. I thank Lorelle Semley, my mentor for the Woodrow Wilson National Fellowship Foundation Career Enhancement Award in 2018–19.

I have benefited from several scholars who have generously shared their primary sources with me: Thabiti Willis for the transcriptions of James White's CMS papers; Kristin Mann for her notes on Dosunmu's land grants, Deniga's *Notes on Lagos Streets* and the Crown grants from the 1860s and 1870s; Halimat Somotan, who shared with me material from the Herbert Macaulay Papers during the pandemic when so many archives were inaccessible; Tayo Pedro at the National Museum; and Susan Rosenfeld, who shared nineteenth-century Lagos court cases with me.

At the University of California, Riverside, I have found intellectual community and fun with my colleagues Megan Asaka, Xóchitl Chávez, Matt Durham, Cathy Gudis, Juliet Morrison, Jorge Leal, Antoine Lentacker, Claudia Holguin Mendoza, Nawa Sugiyama, Flip Tanedo, Cathy Thomas, Jasmin Young, Fatima Qureshi, and Fariba Zarinebaf. History Department staff, especially Michael Austin and Allison Palmer, while overworked, were critical in supporting complex research trips in the US and abroad. I am grateful for the feedback and support of many scholars, including Andrew Apter, Leo Arriola, Rachel Jean-Baptiste, Corrie Decker, Sylvester Ogbechie, and Muey Saeturn from the wider University of California system.

I have presented sections of this book at several academic organizations I belong to. Many, many thanks to my Association for the Study of the Worldwide African Diaspora (ASWAD), Lagos Studies (LSA), and Nigeria Studies families, who have listened to many variations of this project. Thank you, all. My wonderful friends Adamma Oti, Chike Okonkwo, Nike Taylor, Terence Taylor, as well as Alero Akporiaye and Eno Ebong, have been a brilliant support system. My research collaborators in Lagos, Pẹlu Awofẹsọ and Olalekan Adedeji, were critical in making the urban fieldwork possible in 2018, 2019, and 2022.

I have revised material that was previously published as "Historical Tours of 'New' Lagos," *Comparative Studies of South Asia, Africa and the Middle East* 38, no. 3 (2018): 443–54, and "Mapping Old Lagos: Digital Histories and Maps about the Past," *The Historian* 82, no. 1 (2020): 51–65. They appear in a revised form in the "Note on Method" and in chapters 1 and 5. They are used here with the kind permission of Duke University Press and Routledge.

My family has supported me throughout the journey of this book, even when it was not clear what it was all about. Our now-weekly Zoom calls mean everything to me, since we are scattered in several parts of the world—from Lagos to San Francisco. Everlasting love to my parents, Ọmọṣalewa and Julius, and Carolyn Bryant, my mother-in-law; to my siblings, Olufunkẹ and her husband Akin, Adewale and his wife Shelly, and Adeolu; my niblings, Tamilọrẹ, Ikẹoluwa, Ẹniayọ, Tioluwa, and the ultimate Fiki. As well as Adenya and Adelyn. And to my cousins, Solveig, Gimme, Yọsola, Tayọ, Bukọla, and Olumide.

And gratitude to Jody, without whose love, help and support, none of this would be possible.

All the errors are always my doing.

Introduction

Encounters on Land and Lagoon

> Joe, we have your letter, but regret that it is crowded out of this issue; you forget that Lagos has what England has not, a climate. It is composed of equal parts of sun shine, mosquitoes, sand, high temperature, sweetness, light, sound, fury, rain, rage, retribution, and that sort of thing.
>
> —*Lagos Observer,* April 6, 1882

IN 2010, when I was waiting at the British National Archives in Kew Gardens for my Foreign Office documents to be delivered, I noticed that the archive had a map room tucked away on the top floor. The sketches, maps, and plans I found there changed the direction of my dissertation research and, eventually, this book. Afterward, I found them nearly everywhere. The maps were most plentiful in archives around London, Cambridge, Oxford, Lagos, Ibadan, and New York, and each site held variations of representations of Lagos and the lagoons.[1] I cite at least three dozen maps in this book, which form the basis for imagining the city. In this book is a story that they tell together.

How did the British project their territorial ambitions upon cities on the West African coast? And how can we find and analyze indigenous people, knowledge, and ideas in these maps? *Imagine Lagos* interprets a variety of historical maps as a way of understanding how these people and places were framed in the nineteenth century, before Lagos was colonized, and as Britain's territorial conquest expanded. But even as administrators refashioned these spaces on paper according to their needs, they still retained an element of the local. In this book, I read these maps closely for the levels of geographic, ethnographic, social, and spatial information they leverage.

I have written elsewhere about the argumentative power of maps.[2] Due to the different arguments these maps make, I have divided the maps of Lagos and the lagoons referenced in the project into categories. Thus, they are not featured here chronologically, but according to the arguments they support (meaning that a single map can exist in different categories depending on how it is read). No map rests on a single lie. Instead, they project a type of truth that, at first glance, escapes scrutiny. This is why Lawson's 1885 plan of the city can ostensibly represent religious practice by featuring only Christians, while excluding Muslims and those who worshipped *orișa* (Yoruba deities); or why in John Glover's 1859 map a dozen White Europeans can be featured more prominently than twenty thousand Lagosians, in their own city, no less.

None of these maps were made by ọmọ Eko—that is, Lagosians indigenous to the island—but some, like Abbé Borghero's 1865 map of the Slave Coast, contain information that points indelibly to local help.[3] Inscribed on that map is a message accompanying the path that connected Allada to Porto Novo: "interdit aux blancs" (forbidden to whites). The fact that the path was forbidden to Borghero and his allies did not mean they lacked access to information about how to move between both spaces. Each map relies on a specific data set, and each frames space in a specific way. Each map offers its own conclusions about the ways to see Lagos and the surrounding

FIGURE 0.1. Rev. Samuel Ajayi Crowther showing a map of Lagos to Prince Albert, Queen Victoria, and Lord Wriothesley Russell at Windsor Castle on November 18, 1851. From Herbert Samuel Heclas Macaulay, *Justitia Fiat: The Moral Obligation of the British Government to the House of King Docemo of Lagos; An Open Letter* (London: Printed by St. Clements Press, 1921).

cities, towns, villages, and lagoons. These maps rely on a certain intertextuality, especially when produced by the same authority; in some cases they directly reference each other. Some maps are composite creatures: John Pagan's 1883 map of Lagos traced its shoreline from maps by Lawson and from Glover's 1859 *Sketch of Lagos River.*

Imagine Lagos is a spatial history of old Lagos—that is, the city as it evolved between 1845 and 1872 under the influence of Ọba Kosọkọ, Oṣodi Tapa, and Glover. This period falls between a civil war and the first decade of Lagos being absorbed into the British empire. This book examines maps as repositories of history, and mapmaking as a style of writing urban history. In this case, maps join other texts—including letters, reports, and journals—in creating a complex collage of urban life in Lagos. It offers new ways of writing and mapping urban environmental history in West Africa that can access indigenous conceptions and use of space in the nineteenth century.

Imagine Lagos poses questions about the histories of people and their urban and environmental encounters in Lagos. A range of spatial queries animates this book: How did Lagosians insert themselves into a city where they were being erased from the administrative and religious record? What parts of the city harbor the past, and how can they be found? What role do the contours of the past embossed in city space play in highlighting the past? How should we read the city for evidence? What additional details or perspectives do spatial analyses yield in this context? What is the utility of rebuilding with an incomplete and sometimes fragmentary archive? How did mapmakers render these spaces and people through time, and what sociocultural details did their maps and sketches capture that their written documents did not? By posing and answering questions about people, their island, and their city, this book demonstrates how place, people, and urban transformation are connected and how they changed in tandem through time—through war, enslavement, and colonialism. Marking and understanding the ways these places and meanings changed over time is critical to understanding the histories of Lagos, starting in the mid-nineteenth century, and how the consequences of change resonate in the twenty-first century.

My central claim is methodological; I posit that careful attention to space and spatial practice—even when the evidence is fragmentary, weak, or contradictory—helps us to understand historical change, especially in sites and cities where oral cultures have been overwritten by time, infrastructure, and colonialism. For Lagos, these claims are borne out of a careful reading of the available evidence: In addition to the typical archive of letters, oral narratives, and long-held traditions, I analyze the maps of the city, and the evidence stored in the city itself, from the urban fabric of Lagos Island to the geographical context of the space. I establish a method of "walking cartography" to collect, visualize, and analyze this data in new maps of the city. As such, I argue that building and reading this dynamic collage—as it represents the dense overlapping edges and interior of the city and island—gives unparalleled access to the history of the city.

Thus, at the center of this book is the question of how to reconstruct the built environment within this lagoon community, where people who, despite their rich cultural expressiveness in song, dance, and narrative, were often nonliterate, and then, built with the semi-permanent materials that reflected their geography. To do so, every chapter reconstructs a key episode in Lagos's history, focusing on the ways the event affected people and the places they inhabited. Each chapter begins with setting the geographical context and then describing the built environment. When knit together, these physical landscapes provide a platform for the narration of social landscapes, bringing in protagonists who otherwise have no sustained context or enduring voice of their own.

Each chapter is connected to a specific intersection (named in the subtitle) on Lagos Island; for instance, the chapter on the British bombardments of Lagos (which consisted of two attacks in November and December 1851) begins at the junction of Ẹlẹgbata and Lake Streets (where the creek saved Olowogbowo from being completely destroyed by fire). This project is also a cartographic narrative that illustrates the ways the city is used: in each chapter, I draw a new series of speculative maps of Lagos and the lagoons. The new maps in *Imagine Lagos* are plotted to provoke questions, rather than simply settle debates about the contours of the built environment. So they are interwoven into the texts, not merely for illustrative purposes, but to continue the text-based narratives that animate the project. These new maps that I have created are hybrids, giving an effect of reconciling metaphors of language, space, and representation.

Imagining Lagos also means reconstructing the lives of ordinary inhabitants of the city. Many of these people were nonliterate, and dependent on the oral transmissions of their histories, cultures, and personal narratives. Few of these narratives survived the waves of destruction of the city—whether caused by war, civil strife, or enslavement—and even fewer entered the traditional archival record created by consuls, governors, cartographers, and Christian missionaries. Wherever glimpses of these people have survived, the template of a rebuilt city allows an anchor with which to set their stories, however fragmentary they may be in their existence.

While social histories emphasize that the "who" and "when" matter, spatial histories insist that the "where" is significant as well. Before British occupation, why is physical Lagos so unrecognizable, so featureless, and so unchanging in historical narratives? Kenyan writer Binyavanga Wainaina noted that "we are made from our archives."[4] Most accounts of mundane, ordinary days in Lagos, if ever there were any, did not make it into the traditional archive. Lagos of the archive is rendered most visibly as a colonial creation. Land in Lagos has always had important communal, cultural, and symbolic power, but by the 1850s, it also gained more economic value for individuals and their families. Historians have often thought of this land in terms of its value and ownership, rather than for where it is located or its quality. This idea of land as something to be cleared, acquired, divided, mortgaged, sold, willed, or inherited has had a profound effect on how the city is understood.

Imagining this city inevitably creates gaps because this vision is based on an incomplete and often fragmented set of records. While residential quarters like Iga, Faji, Ẹhin Igbẹti, and Olowogbowo are relatively well-documented, we have little to no historical information about districts like Itọlọ and Ọfin. Most of the information we have for the mid-nineteenth century is around Isalẹ Eko (especially the areas around the palace), the Marina, Brazilian Town, and Olowogbowo.

Lagos's urban fabric is as uneven as any old city that is a meeting point for different regimes of power (see figure 0.2). Nowadays, the city's fabric is a typical postcolonial urban palimpsest, with clues to its historical contours, places, and sites peeking out. A street's sudden turn or curious dead end makes sense when read through the shapes of places that have disappeared. The curves of streets follow the former boundaries of lagoons and old neighborhoods like Iga, Ọfin, and Ereko, an area of western Lagos

FIGURE 0.2. Colonial Lagos, from the roof of the Court House. Courtesy of the National Archives, Kew, UK.

where "slum clearance" practices hollowed out layers of the past. On the southern edge of the island, the straighter and wider streets point to the edges of the British settlement.

Narratives that position Lagos as an inherently dangerous and savage place date back to the 1780s, when European traveler John Adams reported seeing decapitated bodies, impaled as sacrifices to Olokun, the water deity, lining the city's shore.[5] Later, from the late 1840s to 1851, most of the information on Lagos came from people writing from outside the city, eager to take over and displace Ọba Kosọkọ, who as a prince was thrice passed over for the throne of Lagos. Lagos was, and is still, a challenging place to live. Despite the fortuity of its location, and the relative safety of living on an island detached from the mainland, residents still had to deal with shallow malaria-ridden mangroves, sickly swamps, and a nearly insurmountable sandbar that made access to the ocean nearly impossible without the risk of capsizing and that claimed many lives and ships, offering them up to the shark-infested waters. In the 1850s, it earned its nickname as the "bug-bear of the Bight."[6] Also, Europeans imagined Lagos as a slave town, an urban space so checkered by the wounds of enslavement that it had to be destroyed and made over. This has meant that scholarship has deferred to the calls and complaints in the sources, neglecting any possibility of redemption. Everything had to be "fixed," from the roads to the buildings, and eventually in 1861, with the annexation, the people themselves. There is a tension between understanding and glamorizing a place, while acknowledging the challenges the residents continue to face. In writing about Maputo, for instance, David Morton shows how in pathologizing a space, "we risk," in his words, "turning these spaces into mere abstractions and dehumanizing the people that live there."[7] Analyzing local responses to "improvement" schemes reveals the symbolic value people assigned to spaces.

Each map represents snapshots of encounters between the local population and the European cartographers who surveyed the spaces. Some maps studied for this book, such as the 1891 *Plan of the Town of Lagos,* were drawn at a scale that lets them be studied as texts. The level of detail in these maps allows them to be read and georeferenced onto contemporary maps of Lagos. These maps cannot and did not intend to present space with any indigenous understanding, and thus any conclusions drawn are already limited by colonial geography. Even if these maps offer dramatic access to the layers of the city, there are still, as Vincent Brown has recognized, "entire worlds that they simply cannot convey."[8] These maps will never fully show the sights, smells, sounds, and other bric-a-brac that make up the city.

At the national archives in Kew and Ibadan, correspondence on Lagos begins in the Foreign Office files, and then moves to the Colonial Office after the city's annexation. I read the letters of consuls and governors and administrators to access the administrative routines (and day-to-day politicking) involved in the running of the fledgling consulate, then colony. In their letters are articulations of intertwined racial and spatial projects as they grapple with the limits of a weak and underfunded apparatus, always vulnerable to the changing tensions in the interior. Rare finds in these archives include records of emancipated peoples. It is unusual to find first-hand documents produced by non-elite figures. Rarer still are testimonies of enslaved people who were drawn to Lagos in pursuit of freedom.

Young women are more visible in the records than older ones, sometimes because visitors recorded what they observed them doing in the streets, whether they were cooking, trading, applying makeup, or even being attacked by both local and foreign men. We rarely find evidence of ordinary days, instead most of what is recorded is the nonquotidian, extraordinary cases that made it into newspapers, into courtrooms, and onto maps.

Land disputes have also generated an important data set for understanding the city. They were frequent in Lagos and were often resolved informally between the parties concerned, or in court. I analyze these disputes together with the fix-and-fill and perimeter maps of Lagos to establish as accurately as possible the locations of historical sites and individual Lagosians who lived in the city, especially before the bombardment.

These land disputes are useful for corroborating historical information and claims made in legal cases—for example, the foreshore cases involving claims against the government by John Holt and W. B. McIver and others. In another case during the 1860s, Banner Brothers (the eponymous company of British owners of commercial warehouses in Lagos and Badagry) had a dispute with the colonial administration over the ownership and distribution of the land in Ẹhin Igbẹti.[9] In chapter 1, I use the written and visual evidence from this long dispute to define the redevelopment of that quarter and the consequences of road building in Lagos.[10]

Even though several of the cases that ended up in court happened after the period the book covers, they are relevant because the historical claims made during testimonies stretch back to the founding of the city, and sometimes address long possession, enslavement, and identity. For information on the early history of Olowogbowo, I read the foreshore cases—*Attorney General v. John Holt* and *Attorney General v. W. B. McIver* decided in the Full Court in 1911.[11] In these two cases, two commercial houses—McIver and Holt—tied possession of their land to the original owners, creating a record of ownership through grants from Ọba Dosunmu and the Crown to before the bombardment of Lagos. The introduction to the summary of cases notes that the Holt case was perhaps the best overview of the history of Lagos available as of 1916.[12] These details are corroborated by the plans of Lagos from 1885, 1891, and 1908.[13] In chapter 3, I use Consul Benjamin Campbell's conflict with the agents of the Church Missionary Society to corroborate these early ownership claims. Even though their struggle was among themselves, their letters and Consul Campbell's intimate map (see figure 3.1) of the conflict also mention Lagosians who were displaced because of the bombardments in November and December 1851 and in the land grabs that occurred afterward. In chapter 1, I analyze the case of *Inasa and Others v. Chief Sakariyawo Oshodi* to establish part of the founding of the Ẹpẹtẹdo district, and the etymology of some street names in Ẹpẹtẹdo in the 1860s. I use *D. W. Lewis and Others v. Bankole and Others* in chapter 1 to establish Tom Mabinuori's long claim to land in Olowogbowo.[14]

The image of Lagos that emerges from the sources, starting with the rubble and ruins from the December bombardment, is at odds with itself: it is fuzzy, incomplete, and often contradictory. The effect is an uneven urban space superimposed on an island filled with swamps, creeks, small peaks, and dramatic slopes, framed by the shallow Ọsa, or Lagos Lagoon. Southwest of Eko, Kuramo Island (now Victoria Island) is separated from Lagos by the Odo Alarun, better known as the Five Cowrie Creek. South of this was the treacherous sandbar.

We need to develop a new language for thinking about cities on the West African coast, and we need to create a visual style for expressing how urban change manifested in the nineteenth century. Combining digital methods with spatial history offers opportunities to do both.

WRITING THE NINETEENTH-CENTURY CITY IN WEST AFRICA

Imagine Lagos builds on a large body of scholarly work that perceives the city as a site of inspiration and renewal and as an ever-complicating puzzle. Writing a history of nineteenth-century Lagos rooted in space, the city, and its population is decidedly different from the scholarly work dominated by the divide between studies of slavery and studies of infrastructure, colonialism, and underdevelopment in the twentieth century.

This book makes two specific contributions to the historiography of cities in nineteenth-century West Africa. First, I situate the history of Lagos and Lagosians in the bounded world of the Benin lagoons. Set between the

Atlantic world and the Yoruba hinterlands, Lagos drew much of its economic and political power from the lagoons that framed the cities to its east and west. As historians have shown, these bounded ecosystems proliferate all over West Africa, but few scholars have taken up their potential for framing the past. As a spatial history, the book takes cues from the research on environmental histories, pushing this discourse east into the mangroves, swamps, and shallow lagoons that frame the cities.[15]

The second major contribution of this book is methodological, in the ways it combines mapmaking and walking as a strategy to reconstruct and analyze the history of the old city. In this book, I have created new maps and texts to narrate a history of Lagos life. By connecting mapmaking with historical narratives, I offer a new methodological approach to the history of West African cities in the nineteenth century. I call this mapmaking "speculative" because there is no fixity in the results of layering new information. Every line, each color, every erasure represent a risk I have taken in narrating the past. Further, evidence continues to accumulate well into the twentieth century about who had claims to the city and what they owned.

Each new map represents a dialogue with the past and the present. Historical spatial data are not as effective without the context of Lagos Island. Each new analytical map responds to the book's spatial queries about history, place, and archiving the past with data that determine the spatial consequences of new political, economic, and cultural actions like war, colonialism, and commerce—including, tragically, the buying and selling of people.

I constructed the analytical maps with a view of the present because that is the way we interpret history: we must always consider the present, try to imagine the past, and avoid the teleological impulse that comes from already knowing the outcomes. Each map comprises new historical and spatial data layered above a contemporary satellite map of Lagos Island. The data for each map is collected from a range of sources, including photographs, maps (e.g., see figure 0.3), letters, and newspaper articles. The sources for each chapter are summarized visually in a cartouche that represents the kinds of material integrated into the layers. See "Dots and Lines on a Map: A Note on Method," which follows this introduction, for more detail.

A focal point of the book is the British bombardment of the island. Focusing on the bombardments addresses an old but foundational question about the history of Lagos. The hostilities that broke out on November 25 and around Christmas Day in 1851 marked a critical point in Yoruba-British

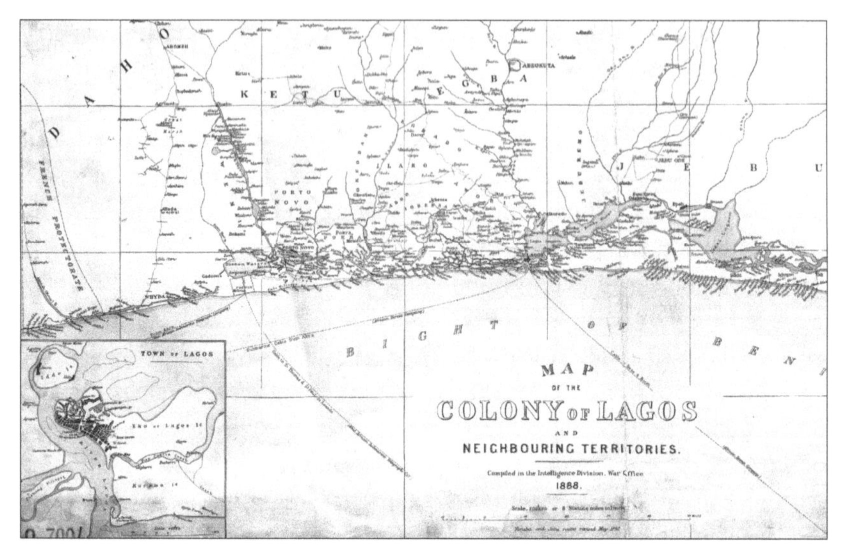

FIGURE 0.3. Detail from *Map of the Colony of Lagos and Neighbouring Territories*, 1888, 1 inch to 8 statute miles (scale), compiled in the Intelligence Division, War Office, CO 700/Lagos 15, courtesy of the National Archives, Kew, UK.

relations on the coast and have evolved into a symbol of British intervention and the conquest of the Yoruba in modern historiography. The British were successful in establishing formal control on the island, a precursor to the colonial rule that eventually extended deep into the mainland. Scholars remain divided on the motivations of the British in these circumstances. There is silence about the motives of the Yoruba, who are usually portrayed as the victims of British aggression, and the violence that was carried out by both sides is usually framed as an "attack" by the British.

Prior to Nigeria's independence from the British in 1960, most histories of the former colony were written by British historians. Although the written primary sources focus mainly on the British side and are sometimes ambiguous, British scholars, such as William M. N. Geary, Alan Burns, and James Smoot Coleman, argued that these violent interventions (the blockades, bombardments, and deposition of the *ọba*, that is, the king) were the inevitable result of Ọba Kosọkọ's unwillingness to give up the slave trade.[16] But Nigerian scholars questioned this rendering. For example, Jacob Ade Ajayi showed that this unnecessarily simplified "moral" view reduced the complexity of the historiography.[17] Using their expertise to render the local point of view, these Nigerian historians attempted to recuperate other sides of the story, to move away from a one-sided discourse on the issue. In this text, however, the bombardment is the starting point for several interventions around space, place, and understanding cities and city life.

These violent acts were not unusual in the region. The early nineteenth century in the Bight of Benin—an approximately four-hundred-mile stretch of coastline beginning at Cape Paul in present-day Ghana, spanning the coasts of Togo and the Republic of Benin, and ending at the outlet of the Niger River in contemporary Nigeria—was a time of urban renewal through methods that were often destructive. In the 1843 edition of his *Vocabulary of the Yoruba Language,* Rev. Samuel Ajayi Crowther included, as an example, a Yoruba proverb about war: "Ogun fọ illu na yan yan" (War destroys the town entirely).[18] This was definitely the case in many parts of Yoruba land, especially with the destruction of the Ọyọ empire in the 1820s, and later of Lagos, twice since 1845. Some of the most dramatic examples of urban transformation in Lagos were tied to political change. In June 1845, a civil war between rival groups meant that the city was sacrificed, and when Kosọkọ challenged Akitoye, the incumbent king of Lagos, the act of destroying several quarters of the city was critical to his winning the throne.

Internecine wars destroyed vast numbers of towns farther north, with the Ẹgba alone losing 130 cities.[19] But new cities came as a result of this destruction, and others were reinvigorated. Ibadan was formed around 1829 by migrants fleeing war, while in the southeast Ẹgba refugees gathered for safety near the Ogun River, close to the base of a massive rock formation on its western bank. Their city, built beside these stones, came to be called "Abẹokuta." The river was an important conduit to the coast. Ninety miles downstream, the river connected to the Benin lagoons that flow parallel to the Atlantic Ocean but are protected by a narrow strip of land. Badagry, Porto Novo, Ẹpẹ, and Ouidah were all older, smaller cities framed by these lagoons. A few more miles south, the river led directly to the only natural opening on the Atlantic coast. It is precisely at this opening that the city of Lagos was founded on three islands: Eko, Ido, and Kuramo. Christian missionaries regularly insisted that the future of the Bight of Benin lay at this entrance of the Atlantic coast, and this entryway was precisely where Lagos was located. As the only "natural" harbor along the coast, Lagos evolved into the focal point of this linked set of cities.

These five cities—Abẹokuta, Badagry, Porto Novo, Ẹpẹ, and Ouidah—share the Benin lagoons and the Ogun River. The conceptual and environmental framework for this book is the extraordinary sequence of lagoons, lakes, rivers, and land routes that connect a series of cities, towns, and villages on the Bight.[20] Individually, lakes like Nokoue can be seen in service of a single polity, but together they form a mostly uninterrupted route hugging the Atlantic coast, conducive to communication and relationship building. These areas shared the same currency, with cowry shells functioning as the means of exchange from Popo to the Kingdom of Benin. Yoruba and Popo were the prevailing languages. The historiography of the Bight of Benin has acknowledged these routes but has hardly leveraged their importance to frame or narrate historical events.[21]

By gaining a foothold in Lagos, the British could access its interior, and they eventually expanded their territory hundreds of miles north, forming colonies and protectorates that came to be Northern and Southern Nigeria, and eventually Nigeria. This trajectory has pushed a teleology onto the study of Lagos as the future capital of Nigeria and the beginning of the story of the country. This book looks at connections that were of the most significance in the mid-nineteenth century. Lagos was indeed a gateway, but its most immediate ties were to the towns and villages connected to it by the Benin lagoons and by the Ogun River. Relationships in the region call to mind an inverted *T* shape. Porto Novo and Ẹpẹ are the farthest points (in the east and west, respectively) in this system, linked by the lagoons that run nearly parallel to the Atlantic Ocean. Abẹokuta is the northern vertex, linked to the Atlantic by the Ogun River. Lagos is at the junction of the river and land routes, thus offering possibilities for cultural exchange, politics, and trade.

MAPPING LAGOS

While most maps typically make arguments about their present, several maps of Lagos also make arguments about the past. For instance, the 1891 town survey marks the boundaries between different quarters in Lagos, allowing one to speculate about the possible extent of some of the historical quarters that emerge but that are not part of the visual record before colonialism, such as Idunmọta or Idunṣagbe.

European maps are an important source for the histories of coastal cities like Lagos. Beginning in 1826, naval authorities scanning the coast drew a variety of maps. But many of these maps were drawn outside the city. It was not until the 1850s that the city emerged more clearly on maps. By this time, several maps of the West African coast, and of Lagos and adjacent territories, had been generated by various sources, such as missions for European institutions and governments, explorers, and missionaries of various denominations. In this book, historical maps become a starting point for the analysis, exploration, and changes in the social fabric of urban life in Lagos.[22] Unlike in India, where there were extensive trigonometric surveys in the eighteenth and nineteenth centuries, serious cartographic interventions in the Lagos region were few and far between. The diagrams of Lagos and other cities in the Bight of Benin represented piecemeal snapshots of a lagoon community that stretched over four hundred miles of creeks, rivers, and lagoons on the Atlantic coast of West Africa.

However, these maps cannot be taken simply as they appear. Brian Harley, a historical geographer, was one of the first scholars to draw attention to maps and mapmaking as critical sites of inquiry for historians. In his landmark 1989 article, "Deconstructing the Map," he questioned the idea that maps have a straightforward relationship with the reality they purport to represent. Harley pointed to the ways in which mapmaking is a representational project, with silences and omissions that are just as powerful and significant as the features cartographers choose to represent. In other words, maps are "cultural texts" that, like other visual sources, are socially constructed interpretations of reality that are "argumentative in orientation" and therefore specifically "propositional by nature." Since power is "exerted on" and "with cartography,"[23] it is important that scholars pay heed to the ways maps have made and continue to make arguments about Africa and Africans. And how we use them as part of our teaching tools in the classroom.

Historians have pointed out that what maps mean "depends on what we think they are."[24] What kinds of information do we get from the maps of Lagos? The maps offer three kinds of data about the past. First, in some cases, they corroborate what the traditional oral account asserts: for instance, that there were four or five original residential quarters in Lagos. Second, they also contradict parts of this record. And, finally, they provide new or forgotten evidence about the ways space was used in the city. Here, topographic and urban data are particularly useful. For example, at some point in the late nineteenth century, there was a triumphal arch built at the intersection of Ẹnu Ọwa and King Streets. The location for this "King's Arch" was next to the highest elevation in the Isalẹ Eko district, which was probably the site of major processions and festivals, implying that the approach to the Iga Idunganran (ọba's palace) would follow the gentle slope down King Street. However, there are some things maps cannot capture, such as every nuance in indigenous Yoruba ideas, or even the most obvious use of terms. For instance, the ọmọ Eko refer to the quarters around the palace as Isalẹ Eko. But I have not come across the name Isalẹ Eko in any historical records before the 1870s.

In this book, I carefully analyze several maps made after 1845. Some of these maps explicitly store information from the nineteenth century but must be read carefully because they represented their observations about Lagos and Lagosians. Two cartographers loom large here: John Glover and William T. G. Lawson, as they produced some of the most compelling maps of Lagos in the nineteenth century.

Maps have been analyzed as important sources for understanding national and urban histories, but mostly outside the African continent.[25] This book also brings African cities into scholarly debates about the importance of maps and mapmaking, especially how cartographic history is a part of an imperial toolkit for shaping and claiming dominance over much of the global south. However, most of these texts focus on India and Latin America. For Africa, Julie MacArthur's *Cartography and the Political Imagination* draws on twentieth-century political struggles to show how mapmaking was used by local people to make claims about identity during the decolonization of Kenya. *Imagine Lagos* brings this conversation to nineteenth-century West Africa, to a region that has often been imagined as unremarkable and absent of any large discoveries by European explorers.

Every *meaningful* map constructs an elaborate ruse; *excellent* ones make that lie invisible.[26] So although maps are useful as historical sources, understanding their arguments and obfuscations is critical for using them to study the past. Even the best-known and most reproduced map of Lagos—Lawson's 1885 *Plan of the Town of Lagos* (figure 0.4)—offers a specific and misleading rendering of the city.[27] Lawson's map highlights the growing influence of British planning on the city. It was first displayed at the Colonial and Indian Exhibition in London in 1886. In its review of the Lagos pavilion, the *Times* reported that "Lagos looked well *on paper,*" hinting even then that the real city might have more in common with the rudimentary "native" objects surrounding it.[28] However, Lawson's map is particularly persuasive in that, at first, it convinces the viewer there are few questions to ask about its composition. This map is the image of nineteenth-century Lagos most often reproduced in the city's historiography, yet few realize its level of distortion in orientation, population, and composition. With its straight lines, sharp edges, and neatly outlined rectangular plots, this map suggests a reading and rendering of the city that belies the tensions in the original patterns of settlement and expansion. The city is stripped of cultural markers that might "threaten" its fragile modernity; there are no Muslims, areas for traditional worship, or even local chiefs in this Lagos.

I read the maps of the coast, Bight, and the interior as texts that frame the historical understandings of these spaces. As Karin Barber writes, "Though many people think of 'texts' as referring exclusively to written words, this is not what confers textuality. Rather, what does is the quality of being joined

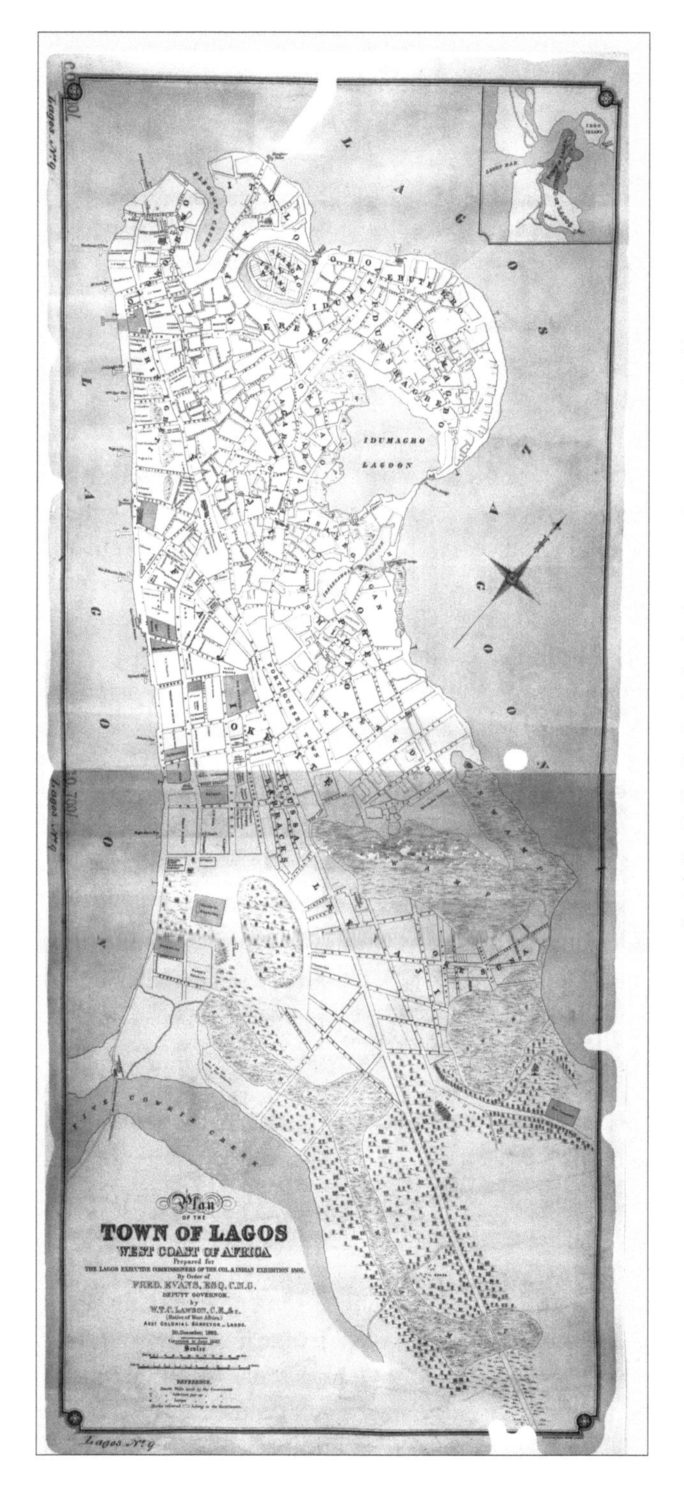

FIGURE 0.4. W. T. G. Lawson, *Plan of the Town of Lagos, West Coast of Africa: Prepared for Lagos Executive Commissioners of the Colonial and Indian Exhibition 1886; By Order of Fred. Evans, Esq. CMG. Deputy Governor,* 1885, about 1 inch to 275 feet (scale), CO 700/Lagos9/1, courtesy of the National Archives, Kew, UK.

together and given a recognizable existence as a form."[29] Therefore, as texts they can be read and analyzed as carefully as any letter, report, or account book. Through an analysis of space, the environment, and people within the same framework, a new history of Lagos and Lagosians is possible. I have divided the forty-four historical maps this book uses into eight categories (discussed in the next section, "Dots and Lines on a Map: A Note on Method"), which are based on the kinds of spatial arguments they make.[30]

Like photographs, paintings, and other visual materials, maps are becoming increasingly important in shaping and answering questions about space in cities like Lagos. The sketches, maps, surveys and charts that sought to represent various parts of the Bight of Benin are an invaluable, if underused, source for analyzing spatial relationships in the nineteenth century. Maps made the world legible, but only according to the priorities of the cartographer. They were part of the imperial arsenal, as valuable to them as the gunboats, treaties, and officers who created and wielded them. Inadvertently, these maps that aimed to obscure the spaces where Africans lived resulted in preserving them in time. Also, the correspondence of colonial officers and missionaries, produced around land disputes, bombardments, and development of new infrastructure, helped frame the ways indigenous people lived. At no point is Lagos more visible than when it is marked for destruction, division, or "civilization"; thus, I use these maps, documents, and indigenous sources to reconstruct the past, in place.[31]

In writing about cities, Orhan Pamuk tells us that "what gives a city its special character is not just its topography or its buildings, but rather the sum total of every chance encounter, every memory, letter, colour and image jostling its inhabitants' crowded memories."[32] While social histories of the city have taken up this challenge, dynamic maps add another useful layer to the interpretation of space, and in the plotting of patterns that are only obvious when seen in relation to each other.

A map is rarely a stable or final text.

Indigenous voices crept into these maps at unexpected times: to help navigate a dangerous turn in a river, or even to point out a path Europeans were banned from using. Some maps, like the 1865 *Sketch of Lagos and the Adjacent Country,* explicitly reference the relationship between mapmakers and their informants. In its legend, the cartographer mentions that the data for plotting the coastline and lagoons comes from the earlier surveys; however, he notes that the "position of the native towns and the boundaries of their countries are laid down as nearly as they can be ascertained from inquiry."[33] This simple annotation, while acknowledging the importance of local knowledge, immediately posits it as incomplete and implies its potential inaccuracy.

In her writing on Indian cities, Ananya Roy invokes the idea of "unmapping" to describe the process in which neglect becomes an intentional strategy in urban planning and where places are left undocumented in maps and plans as a way of trapping them for future erasure or exploitation.[34] I build on this term to think about how these maps left out the people in the areas covered in them, beginning in Glover's *Sketch of Lagos River,* on which he recorded the dozen or so Europeans who lived adjacent to a city of twenty thousand Africans. Maps of Lagos continuously and consciously omitted these people while mentioning in detail the institutions that served the colonial vision of the city, such as the churches, prisons, racecourses, and courts. While they are considered an elaborate archive of city (and regional) life, these maps reveal why community ties were durable in the ways they persisted; how proximity shaped and sharpened antagonism between people who shared cultural and religious mores but were not of the same race; and, finally, how the city was inverted and ruined as a result of British intervention. The back became the front, driving the local rulers to the edges.

I have identified eight categories of historical maps for this book. I have consulted these maps—from perimeter maps to intimate maps—to glean information about mid-nineteenth-century Lagos. I also drew three new kinds of maps of Lagos to frame the walking cartography of the city. Each of these maps is based on historical maps georeferenced in ArcGIS, analyzed alongside sketches, diagrams, and coastal views. First are the Spyglass maps, which allow you to see the historical parts of Lagos. Second are the infographic maps, which contain annotations, graphics, and spatial interpretation and represent the middle of the mapmaking process. See, for example, figures 0.5 and 0.6. The third maps are the largest and richest in detail and represent the result of my historical and spatial analysis of old Lagos, and are a launching pad for every argument in *Imagine Lagos.* These speculative maps are a combination of the data gathered from the research walks on Lagos Island and the material in the traditional documentary archive. I discuss more technological detail about the making of these maps in "Dots and Lines on a Map: A Note on Method."

DEFINING A WALKING CARTOGRAPHY

The idea of a walking cartography rests on the hypothesis that a city's urban fabric can function as an archive of its own past, despite or because of human and environmental agency and action. This is true in a small, densely populated, and compact island like Eko, where even though change is frequent, it is still limited by space, resources, and human needs.

Creating a walking cartography first requires extensive archival research. It then combines two things: (1) walking the city's streets as a method of understanding and collecting historical data, and (2) mapmaking as a method of storing, exploring, and analyzing spatial data and patterns and of presenting historical data or analysis. Walking is methodologically important for understanding the spatial history of the city and island and for analyzing, narrating, and drawing new maps of the city. When combined with mapmaking, this walking cartography makes for an effective method of reconstructing the past. Historical sites or spaces in Lagos tend to endure, and in some examples, even when they disappear, their meanings persist. Or are transformed into new spaces.

I discovered, for instance, that Lagos's street names anchor the city's histories even when historical spaces disappear, are repurposed, or are written over entirely. To reconstruct the city in 1851, I first analyzed British Foreign Office correspondence from the bombardment of Lagos and the maps produced by naval officers and details about the renaming of different sites and features in Lagos and Victoria Island.[35] Yet there were still gaps in my understanding of the pre-1851 city. Walking in Olowogbowo, Alakoro, and Ẹlẹgbata—sites that were surveyed and partially destroyed—and comparing them to the archival data revealed patterns of confinement and slavery in Lagos and provided important spatial context for the reorientation of the city post-1851. Map 2.2 presents cartographic evidence of the city shifting its focus to the Atlantic shore, the scars from slavery, and the ways that Ọba Akitoye routed his return to power.

I completed this book during the height of a pandemic and policing crisis in Nigeria, the United States, and most of the world. This affected how I thought about spatial history in Lagos, especially with regards to the encounters that ordinary Lagosians had in these new and renewed spaces. In Lagos the #EndSARS crisis made the plight of Lagosians acute and public. While there was obviously no social media in the 1860s and 1870s, reports in newspapers helped to illuminate the goings-on in Lagos while the colonial administration was making twinned claims about improving space and people.

The discoveries during my walking research in Lagos Island led to the realization that streets were key to accessing the past in the city, especially in the lesser-documented residential quarters—such as Idunmọta, Faji, and Ereko—that were so often subject to the tragedies of slum clearance after the 1880s. After I started thinking of histories and street names in cities, I found this idea everywhere. Bookstores in many major cities—at least the walkable ones—have a section or at least a shelf dedicated to historic walks. From Philadelphia to London, New Orleans, and of course New York, writers and scholars have tried to reconcile the relationship between a city's streets and its history. In my interpretation of Lagos's history, I tie this process to cartography. Thus, walking the streets of Lagos Island became a feature of drawing the final versions of the maps.

What traces do people leave in the places where they live, and even where they die? Originally, my research on the history of Lagos Island took me to archives in Lagos, Ibadan, London, Liverpool, and New York. The documents and maps I found in these sites produced interesting, yet strangely static, versions of the city. These often divided the city's history too neatly into precolonial, colonial, and postcolonial contexts. The historical maps did something similar with chronology, but their intertextuality was more promiscuous about time. Some historical maps referenced and superseded each other and paid no specific attention to time. Glover's map of the Lagos River, for instance, may be dated 1859, but it pulls in material mapped in the 1840s and notes written by other people at different times, before and after it was produced.[36] Read and analyzed together, these maps were still missing something, especially when the speculative maps I was able to produce did not resonate with Lagosians in the city.

A chance encounter with the Spyglass application led me to making more dynamic maps. Using ArcGIS, I could layer two maps: the top layer was the contemporary satellite map of the city, and the bottom layer was a geographically correct historical layer. Using an 1891 fix-and-fill map of Lagos as the bottom layer, I could "move" the Spyglass and see historical space right through contemporary places in Lagos.[37] This application proved excellent for tracking some geographical features, like Isalẹgangan Lagoon and Alakoro Island, but it was less useful for the older quarters, like Isalẹ Eko, Idunmọta, and Ọfin, because there are few historical maps that are geographically correct before 1891.

Dissatisfied with this aerial view, and weary of trying to connect mapmaking to only written sources, I ventured into the city to see what traces of the past could be found. I spent three summers walking in Lagos Island—in quarters like Alakoro, Ẹlẹgbata, Oke Popo, and Faji—trying to reconcile the archival and cartographic record with the actual evidence in the city. Initially, the search was for obscure historical elements of Lagos that were mentioned in maps, of which there was little to no evidence of in the archival documents (and vice versa). Examples include the use of "Iga" rather than "Isalẹ Eko" to refer to the ọba's quarter in the northwest; land grants and documents mentioned giving land in "upper Fadge" (read Faji), "the Fields," and other spatial references I could not make sense of. In one clear case, I went looking for traces of the steam tramway that existed in the early twentieth century but seems to have been entirely forgotten now.[38]

Walking in Lagos Island offers no guarantees that you will find any evidence of the past. At first, the shock of the present nearly entirely obscures the historical detail in the streets. It can even feel like history has no space here. Historical research feels like an extravagance given the decaying infrastructure, lack of sidewalks, lack of support for everyday needs, the general din and slowness of traffic, and flooding and trauma everywhere on the island.

In many ways, the intersections and junctions in Lagos function like the monuments and markers that this nineteenth-century city did not have. By coming together and marking space, they hold meanings and memories that residents could write over, project, or erase. Walking, as a form of urban research in Lagos Island, provided significant insights into some of the claims in the archival records of the city. It also provided important information on scale, distance, and density in the city. Between one district and another, there were changes in scale, streetscape, and environment that clearly contained echoes of the past. In Massey Square, for instance, the Marina area felt like a distant memory, and the lagoons that are such an important feature of Lagos's identity did not feel close.

Walking cartography is also a method of trial and error. Just as a map can never be a full representation of a space, walking in the city today can never fully convey a sense of its past. Several itineraries—for instance, walking in the Itọlọ or Ọfin districts—yielded little historical information about the nineteenth century, and some data directly contradicted what I imagined the past of the city to be. However, the city itself bears important clues and remnants of its own past.

Some streets come up frequently in the historical records and anchor most of the discussions in the book. Examples include Marina, Broad Street, and Campbell Street. My research team and I walked on the streets that seemed the most important for making and defining place in the mid-nineteenth-century city: In Olowogbowo, we found our way from Breadfruit Street to the former site of Ita Balogun; in Ẹhin Igbẹti, we explored the echoes of the "three Cs"—commerce, colonialism, and civilization—along Broad Street and Marina Road, which used to be Back Street and Water Street, respectively. In the northwest, we walked around Ebute Ero and the ọba's palace. Historical maps locate the King's Arch on the highest elevation in the northwest, at the intersection of King and Great Bridge Streets. We found little trace of the historical arch. Several routes began from Tinubu Square (or Ita Tinubu, as it was known in old Lagos), a social and spatial fulcrum in the city, and from there we could take paths to many quarters. The northwestern route from there took us through the border of the Ereko and Agarawu quarters via Victoria Road (now Nnamdi Azikiwe). On another occasion, we went east, via Bamgboṣe Street, to the site of the old cemetery.[39]

On these many walks, I identified a variety of themes—from memory to violence and justice—that were key to understanding change over time in the mid-nineteenth-century city. These themes came together to solidify the significance of placemaking itself, and its impact on writing a fresh chronology of Lagos history. In February 1868, a new and far-reaching list of street names emerged in Lagos. This timing is critical as it merged the persistent and historical indigenous symbols in the city with the new infrastructure of British colonial rule.

Each "walking day" began with a set route or itinerary, but had to be flexible enough to incorporate unexpected detours. After collecting the data, I compared it to the historical record. Next, I added them to my satellite maps, and compared them to my hypothesis.

KNOWING LAGOS

Swampy, flat, and sandy, Lagos Island sits low on the Lagos Lagoon on the Bight of Benin, approximately four miles from the Atlantic Ocean. North of the island, the lagoon is shallow, averaging around three feet, but it deepens in the southwest around Apapa, where it reaches a depth of thirty feet. The shallow entrance to the lagoon is obstructed by the dangerous sandbar. The island's perimeter is ringed by a thick belt of mangroves, which features prominently in many of the earliest depictions of the city.

The island's setting juggles many advantages and disadvantages. One advantage is its location at the junction of several trade and travel routes that connect the interior and the coast over land and through river. Some natural disadvantages are the swampy marshes (see figure 0.5) and implacable

FIGURE 0.5. Mangroves around Lagos Island. Photograph by author.

sandbar that separate the island from the ocean. The island is approximately three and a half miles long and one mile across at its widest. In the 1840s and '50s, approximately twenty-two thousand people squeezed along the shoreline into an area less than four square miles. This shoreline is interrupted by several inlets, creeks, and lagoons, in addition to the pockets of swampy marshland, the largest being the Okesuna swamp in the east. The shallow Idunmagbo Lagoon covered around forty acres, and the Isalẹgangan Lagoon, a thin but extensive strip of water, covered around twelve acres. These secondary water features originally limited the habitable area and, in the early nineteenth century, made the eastern part of the island virtually uninhabitable, as it was mostly swamp.

In this book, I use "Eko," "Lagos," and "old Lagos" in specific ways, and which term I use depends on the historical period being discussed. The island has several names, and the names people used often depended on their relationship to the city (see figure 0.6). In his 1877 memoir, one former trader wrote that the town "is sometimes called Eko, but oftener Lagos."[40] But why, and when, does it matter what the city is called? Still today, when older people say they are going *to* Eko, it is widely understood that their destination is Lagos Island, and not the parts of the city that have spread toward Lekki, and across wide swathes of the mainland to the north, west, and east.

As late as 1871, one racialized French space map insisted that Lagos Island and town were still called Eko by the local population.[41] The root of the name *Lagos* is attributed to the Portuguese *lagoa* (lagoon), but even those who spoke Portuguese called the city something else in the nineteenth century: Onim. Although Brazilians and the Portuguese used this name, at least one ọba used it on occasion in his correspondence with the Brazilian slave traders he worked with in the 1840s and up to the bombardment. Eko was a name used locally in the city. To be of the soil was to claim to be ọmọ Eko (a child of the city and, by extension, African/Black), but to be ara Eko was to be an inhabitant, though not necessarily indigenous. British officials called the former "aborigines," and both categories were often lumped together as "native inhabitants" of the city.

EKÒ OR LAGOS ISLAND

FIGURE 0.6. Detail from John H. Glover, *Sketch of Lagos River*, 1859 [corrected to 1889], 1:20 scale, MR Nigeria S.127–24, courtesy of the Royal Geographical Society, London.

Identity in Lagos is complex and layered and has changed over time. Throughout the book, I refer to the African and Black inhabitants of the city as Lagosians. Whenever possible, I make distinctions based on the claims that they made. Some are "ọmọ Eko," in that they make a claim to indigeneity based on a kinship with the firstcomers, usually in connection to Isalẹ

FIGURE 0.7. Detail from William C. Speeding, *Lagos Harbour,* April 1898, 1 inch to 1,600 feet (scale), 63 × 90 cm, courtesy of the National Archives, Kew, UK.

Eko and other older quarters. "Ara Eko" usually refers to Yoruba settlers in Lagos who did not make specific claims to indigeneity. Other Lagosians are of other origins, like the "returnees" from Sierra Leone, Brazil, and Cuba. Some Sierra Leoneans (or Saro, as they were known) were of Yoruba, Igbo, or other origin. All were rescued by the ships of the Royal Navy and settled in Freetown and other parts of Sierra Leone. They came back to Lagos, Badagry, and Abeokuta in search of new opportunities or in search of their families.

ORGANIZATION OF THE BOOK

The book has been designed in a complex, concentric but precise chronology that follows the logic of placemaking in mid-nineteenth-century Lagos. It continues with a note on historical GIS (geographic information system) and mapping Lagos, five chapters, and a conclusion. The chapters are framed around the concept of walking cartography—that is, a series of research outings designed to collect historical and spatial data through the city and island. Each chapter begins at a different intersection in the city and pairs a social and spatial reading of the relationship between people and place in Lagos: social in the sense of reading and interpreting encounters, and spatial in the sense of creating a site to anchor the stories and thus interpret them more fully. Map 0.1 shows the intersection at which each chapter begins.

The first three chapters focus on the theme of risks in representation. Here, representation comes from origin stories, images, memories, and correspondence generated during and immediately after Kosọkọ's reign in the city. Two things are at risk here: first, the making and archiving of this data, and second, the methods of interpreting it. The last two chapters are organized around questions of ownership and belonging in the early years of British colonial rule, when it was still possible to imagine a return to local rule by a royal family whose members had led the city for decades. My analysis of sources around reconstruction point to how the influence of land redistribution and new infrastructure went beyond narratives of improvement, and looks to the ways that belonging was shaped by race, ethnicity, and land in Lagos and the lagoons in the years after formal British occupation. I show how reading a reconstructed city can provide a template for recovering the lives of the city's silenced inhabitants.

In chapter 1, "Streets, Placemaking, and History in Old Lagos," I reconstruct and read the first written set of street names in midcentury Lagos to reveal the ways indigeneity was first marked, and how these names—of

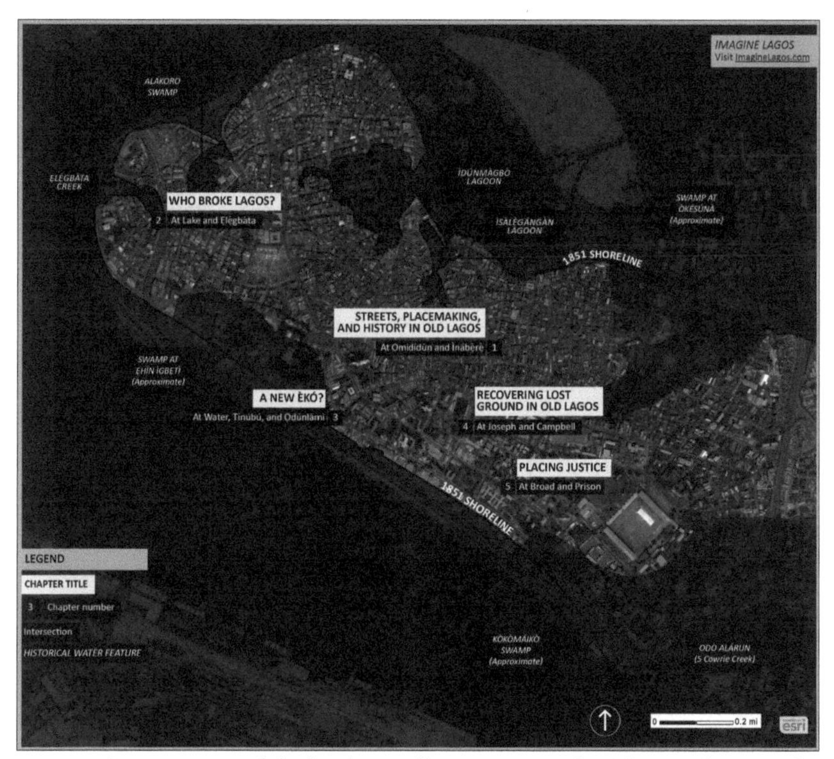

MAP 0.1. Organization of the book: a walking cartography. The numbers on the map indicate the intersection at which each section of the book begins. Base map courtesy of ESRI. Map by author.

FIGURE 0.8. Two cartouches from the *Imagine Lagos* maps.

places, people, landmarks, and the environment—provide a transcript and entryway for understanding space and people in Lagos. A map of the old Marina from 1871 begins the process of linking people and place, and the map I construct in this chapter, "S'a ti Peni sọ adugbo ati ita l'orukọ" (map 1.2), highlights the major streets and generates thematic transcripts that come from reading the names in context. These names, read together and as a whole, earmark districts, protagonists, and, most importantly, paths through the city that lead to protagonists and unexpected encounters in space. By generating the street network in the city and creating a spatial index, chapter 1 allows a spatial context for the makers and landmarks that animate the rest of the book. This chapter is a meta-narrative of space, as understood through the ways in which historical street names marked space in Lagos. In this chapter, I trace my own path through the city, from Ẹlẹgbata Creek to the old Marina, from Alakoro to Ọfin, and from Massey to the Brazilian quarters. The maps and locations spun from these itineraries form the physical basis for the remaining four chapters, which are chronological, and together trace the changes in the urban fabric of the city, from its destruction in the Ija Agidingbi in 1851, through layers of reinvention, and to the end of the first decade of formal colonial rule.

The two chapters that follow concentrate on a reading of the city before the bombardments in late 1851, and the reconstruction of the city after it. Chapter 2, "Who Broke Lagos?," shows how slavery and perceptions of slavery characterized the urban fabric of the city and island. It examines the city in ruins.

"Lagos ought to be knocked down by all means."[42] This was the sentiment expressed by the British prince at the Windsor meeting with Crowther and the Queen (see figure 0.1). It focuses on the time when the city was in fact knocked down. It dwells on two civil wars centered on Lagos (in 1845 and 1851), which functioned as turning points not only for changes in political regimes but also for changes in indigenous spatial practice. By reframing these narratives in maps and texts—as they occurred on land and water—this chapter shows how the Ija Agidingbi, in 1851, was in fact the tail end of unrest spanning Lagos and its lagoons, drawing in a cross section of coastal society from Badagry, Ẹpẹ, and Ouidah to Lagos. In the nineteenth century, Yoruba-speaking peoples settled and built towns and cities in the Bight of Benin. However, little physical evidence of the past exists outside oral traditions passed down through generations, and beneath the uneven settlement of the British colonial city that was built on top of and around old Lagos. Chapter 2 uses texts and archival and cartographic evidence for plotting the old city, within the island's historical shoreline. To imagine Eko, the chapter uses a flipped map of Lagos, one that has the palace at its center and, therefore, the city facing the mainland instead of the Atlantic Ocean. In later chapters, the maps use the north–south orientation that reflects the preoccupation of the mapmakers. The map "Marking Ọba Kosọkọ's Eko" (2.2) focuses on reconstructing the city from the records of the *Papers Relative to the Reduction of Lagos*. In doing so, it addresses the "placelessness" that characterizes so much of the literature on precolonial West African cities and foregrounds the meanings behind the waves of constructing and destroying the city.

The third chapter, "A New Eko?," looks at the process of rebuilding the city after the bombardment and examines the making of a new city, starting from the edges. Almost immediately after the city's partial destruction in 1851, questions of how to rebuild it emerged, prompted by competing claims on space by reenfranchised local elites (mostly the returned ruler Akitoye and his chiefs), newly arrived Europeans, and formerly enslaved people from Freetown, Bahia, and Havana; these questions were all debated within the context of the waning transatlantic slave trade. The new map in this chapter, "A New City Springs from the Breadfruit Trees," focuses on how shaping Lagos was no longer solely in the hands of the indigenous people who returned after the December 1851 bombardment.

In chapter 4, "Recovering Lost Ground in Old Lagos," the book moves inward, zooming in to the residential quarters, working through the vexing idea of specificity without fixity in Lagos and examining the impact of land redistribution in Lagos, from Ọba Akitoye's death to the British annexation of the island. This chapter is based on an important but fragmentary set of sources, Ọba Dosunmu's land grants, which have mostly been hidden from view.[43] The chapter is more experimental as it interprets this finite yet incomplete source base that offers a compelling look into life on the island. Although the source base is incomplete, it is difficult to ignore. As the city began to fill again after the bombardments, land quickly became scarce on the swamp-filled island as newcomers and old residents alike tried to secure space, and what emerged from this process in the consular decade (1851–61) was several land grants offering space in the city. Dozens of people received real plots of land, but the descriptions of the process read like a mystery to be unraveled. In this chapter, I read these grants as a prequel to a series of landmarks—such as the race course, cemetery, and Faji Market—that we see in the 1868 street network of Lagos. Ọba Dosunmu's land grants are an irresistible opportunity to explore a neglected spatial period in Lagos.

This chapter explores the impact of the seventy-six land grants signed by Ọba Dosunmu, the new ruler, and given mostly to strangers. The wording of these grants lent a new grammar to the fixing of space in Lagos, as their language, diction, and intention ushered in ideas of land distribution and thus land organization. Within these grants, we see the gradual emergence of newly legible quarters and districts on the island and in the surrounding region. "To fix" has two meanings in this chapter. First, it refers to the precision in marking off territory as personal—instead of the usual communal—property. Thus, it points to the action of creating new boundaries to maintain and accrue value. Second, fixing refers to the perceived repair of space, as evidenced in the letters, dispatches, and newspaper articles where people frequently remarked on the previous "haphazard," "inconsistent," and "unprofitable" patterns of land ownership and policing of land in Lagos. A new map, "Ọba Dosunmu's Land Grants" (4.2), plots the intertextuality inherent in the language of the grants, in a city where social relationships were key to understanding the physical landscape. By 1853, the flexible customary rules around land tenure began to collide more frequently with foreign ideas about land title and ownership. Whereas in the past custom dictated that land could not be sold, a new system of land grants ushered in a complex process of land redistribution. A system of two towns began to emerge on the island, and we see on the maps and in letters a struggle between whether they remain Eko, become Lagos, or stay as both. This chapter explores these land grants given between 1853 and 1861 to plot the changing relationship of space to power.

The final chapter of the book, "Placing Justice," looks at the first dozen years of colonial rule in Lagos, especially at the claims that governors and administrators like John Glover and Henry Stanhope Freeman made about their improvements in the city. The British government annexed Lagos in August 1861, and this chapter examines the relationships between place, race, and power in early colonial Lagos. The debates were among indigenous Lagosians, returnees from Sierra Leone, and various people of color from within the British empire. The chapter explores their ways of reinventing themselves, their city, and region via their newspapers, petitions, and other publications. This chapter demonstrates how questions of race and identity in an urban context interacted in determining who was local or foreign, and how that factored into the reconstruction of space. It picks up some of the earlier themes around issues of improving the urban fabric. This discourse on improvement itself, plotted both on the ground and at the Colonial Office in London, and based on policies previously implemented in other colonies, masked the utility of existing indigenous urban patterns. In fact, these interventions—the new roads, institutions, and infrastructure—not only made British influence more permanent but also sharpened the possibility of permanent occupation. Map 5.2 illustrates the continued erosion of local power through the permanence of new infrastructure and political centers.

High-resolution full-color versions of all the maps are available online at https://imaginelagos.com.

NOTE ON ORTHOGRAPHY

Since Rev. Samuel Ajayi Crowther translated the Bible from English to Yoruba in the 1840s, there has been a debate on how to properly render Yoruba in its written form. Yoruba, as transcribed in Latin script, employs both lower and upper diacritics. Not all typefaces can reproduce these characters, so for the main text of the book I have only used the lower diacritics—for example, Ẹhin Igbẹti, ọba, and Oṣodi. I have used consistent spellings where they exist and conform to the general rules of Yoruba vocabulary. Yet, inconsistencies remain. These occur because the spellings of words have changed over time or vary in response to regional differences. However, the speculative maps, such as maps 0.3 and 1.2, include Yoruba text written with both upper and lower diacritics, using a typeface that can accommodate them all.

Dots and Lines on a Map

A Note on Method

ALL MAPS should be approached with a certain amount of skepticism, and the ones in this book are no exception, whether they are my own recent creations or were drawn decades, or even hundreds, of years ago. This segue speaks to the process of creating and interpreting the new maps in this book. It should be used as a guide to understand how patterns on the maps of Lagos Island are generated from historical and walking data, and how the interpretation of these patterns forms the basis of each chapter's argument.

Digital methods now allow for incorporating different sources to frame and answer historical questions. In *Imagine Lagos,* I use historical maps as sources and have also created several new analytical maps, using various applications, to shape and support the arguments that I make. This book privileges a spatial interpretation of traditional archival sources, such as letters, treaties, photographs, and land grants. It contains three kinds of new maps of Lagos: experimental maps that focus tightly on individual details (see map 0.4 below); Spyglass maps that allow the viewer to see space at two different historical times (see map 1.1 in chapter 1); and, finally, new speculative maps of places in Lagos Island that feature historical data written on top of a contemporary satellite map of the city (see map 3.2 in chapter 3).

Mapmaking is such a subjective process, yet the results are often so persuasive (and seem so finished) that they can be impervious to questioning.

Selecting what to represent and what to omit produces certain visual outcomes. Each map is a snapshot of a specific set of encounters as static maps are inconvenient for showing change over time. For transparency, I have included these explanations of the methods involved and the choice of sources.

Historical maps hold a lot of data about the spatial arrangements of old cities like Lagos. Yet, maps sacrifice content for legibility. Close readings and textual analysis allow scholars to extract and collect data from maps, sketches, and plans, while mapping software like ArcGIS allows one to plot this data and look for visual patterns and relationships that text alone cannot offer. Further, applications like Adobe Illustrator and InDesign allow the combination of all this material into a visually compelling format. In this section of the book, I show how seven different new maps of Lagos create the spatial context to imagine Lagos, as well as the visual context to continue the analysis based on material in traditional and urban archive.

I had multiple intentions in making the new maps in this book. Most important of which is that they function as a catalog for the kinds of spatial data that can be historicized. In this book, I write about the historiographical blankness of the city's urban past, and how it has been difficult to imagine Lagos in the mid-nineteenth century.

The architecture and infrastructure of the nineteenth- and twentieth-century colonial city are familiar, and so is the architecture of the "returnee": narratives about the first story building in Nigeria and Brazilian architecture in Lagos and the Yoruba interior have been indigenized. This book does something slightly different: it considers the reciprocal relationships of people, their environment, and built architecture in a period I call old Lagos spanning from 1845 through 1872, which marked the end of the first decade of formal occupation by the British.

REIMAGINING OLD LAGOS WITH HISTORICAL GIS

Map 0.2 shows the extent of Lagos Island under consideration in this study. Though ArcGIS and other tools have proved useful in the mapping of London, Paris, New York, and other such well-documented cities, these data-hungry applications are often less compelling in studying African cities, where sources on the precolonial past are scarce and interpreting the written or drawn sources that exist requires a more flexible process. The lack of "standard" empirical data (such as census data) that drives ArcGIS and other such applications is one of the main bottlenecks in using ArcGIS for precolonial West African cities.

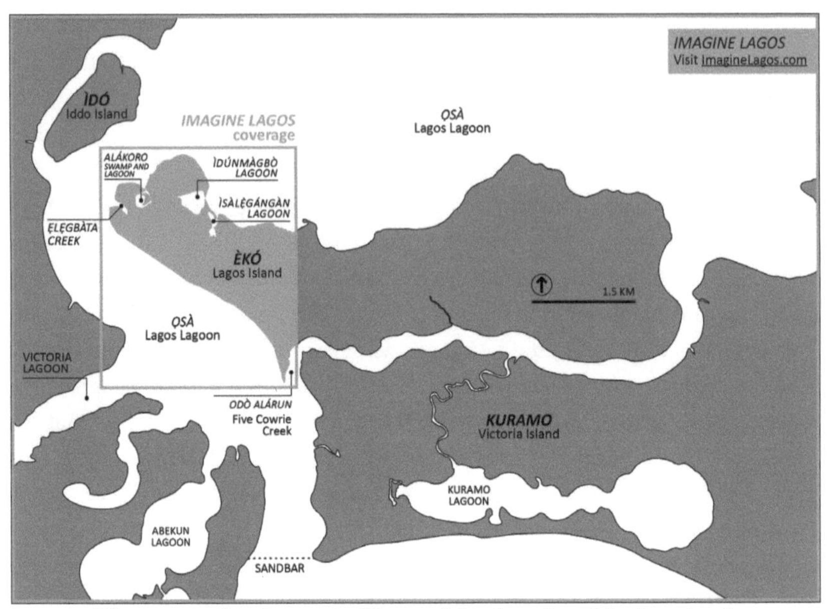

MAP 0.2. The geographical extent of Lagos Island explored in *Imagine Lagos.* Map by author.

The other problem is the method. European visitors to the region produced most of the visual and written sources for the nineteenth century. This means that the sources produced are sometimes hostile to the people and places they seek to represent, and their material is rife with misunderstandings and misrepresentations that range from spelling to factual errors. This analysis thus requires a critical edge rooted in the knowledge collected not only from archives but also from Lagosians and other local sources.

But what kind of city does this interdisciplinary research approach yield? In "Mapping a Slave Revolt," the historian Vincent Brown explains how plotting movements in space allows new insights, and thus the possibility of posing new questions of our sources.[1] What if, as historians, we could compare past (and current) renderings of the urban fabric, and combine this reading with indigenous material such as *oriki* (praise poetry in Yoruba), language, and origin stories? What questions could we ask of spaces that have changed from sites of enslaved labor to churches, or from swamps to markets? With new digital tools such as KnightLab's StoryMap, ArcGIS, MapWarper, and MapScholar, this kind of research is already possible. These maps dwell on the relationship between the various representations of space that are available as sources. But "if a map reflects anything," as Raymond Craib writes, what we see in them "is the relationship between

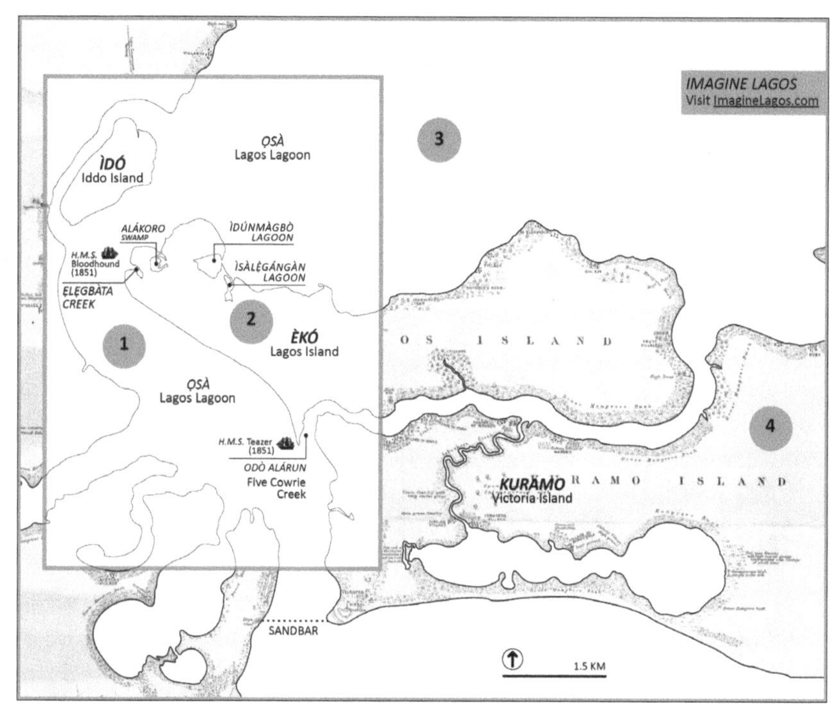

MAP 0.3. "Making Places in Old Lagos." The historical map is a detail from William C. Speeding, *Lagos Harbour,* April 1898, 1 inch to 1,600 feet (scale), 63 × 90 cm, and is courtesy of the National Archives, Kew, UK. The number 1 on the map points to the frame that highlights what part of Lagos Island is the focal point of old Lagos; 2 refers to vector data inscribed on the outline of Lagos Island; 3 represents parts of the map that are removed to maintain the focus on the western portion of Lagos Island; 4 points to the visible section of Victoria (Kuramo) Island from the 1898 base map. Map by author.

modes of representation and the material practices of power."[2] This is all too clear in the maps of the bombardment of Lagos. Even naval officers took it upon themselves to rename parts of Lagos and Ido Island while plotting maps that would help them destroy it. (See map 0.4.)

The maps in *Imagine Lagos* comprise raster and vector data. Here, the raster files are nineteenth-century maps that have been scanned, and often these images do not contain enough spatial data. The vector data are the spatial information, such as roads, rivers, sites, et cetera that have been added to encode specific locations. In order to use these maps with spatial data in contemporary programs like ArcGIS, it is necessary to "align or georeference them to a map coordinate system (MCS)." MCS are developed using map projections, which are defined as the process of displaying the earth, which is curved and three-dimensional, into a platform that is two-dimensional.[3] Georeferencing data is useful in this case because it allows raster data to be studied, queried, compared, or even just looked at with the other spatial data that is available.

When using historical maps for this kind of research, the most important step after digitizing a high-resolution version of the image is georeferencing it. There are several excellent applications for this, and I use ArcMap because it allows for layering and thus comparison of various renderings of Lagos's geography, all relative to the contemporary city. This simple but powerful function allows scholars to acquire and apply real-world data and coordinates to historical maps, providing that the map itself is geographically correct to a reasonable extent.

The best maps for this process will have an abundance of historical markers that have modern-day correspondence, such as important physical features like inlets, lagoons, and streams, or man-made structures that have endured in the same location. Important permanent monuments and buildings can also be used to geolocate and spatialize historical maps. For old Lagos, the earliest maps that work for this process are the ordinance surveys from 1891, and the map of the Lagos Harbor completed in 1898.[4]

For instance, I have traced out the historical shoreline of Lagos (see map 0.4) by comparing the archival record with georeferenced maps over contemporary satellite imagery. Once these maps are georeferenced, they can then be easily manipulated. For instance, in this map of Lagos, on Adobe Illustrator sections can be traced out, drawn on, or even deleted completely.

This process requires control points, which are sites that can be matched in contemporary projects and in the historical maps. Examples of excellent candidates for this process include the 1891 town surveys of the city and the 1898 survey of the Lagos Harbor.[5] Because of their high level of geographical data—the former is an ordinance survey of the streets, while the latter focuses on the contours of the islands—and the endurance of important sites like the British Consulate built in 1853 and the Church Missionary Society (CMS) Church the year before, these kinds of maps are important as background information. After identifying such sites, the historical map is then "transformed" or warped in a process that allows it to "permanently match the map coordinates of the target data."[6] These GIS maps are then processed in drawing programs such as Adobe Illustrator and InDesign, in order to add the data that may not have a specific pinpoint location or may have been reported to be in several sites, so marking it is a best guess or estimate.

Not every map has to be georeferenced, as some spatial data are encoded simply in the interest of clarity.

Each map, though two-dimensional in the book, results from a three-dimensional process of plotting and layering information, which itself was produced and then archived in a context that privileged an external perspective. However, only one topographic layer—that of the satellite map—is visible. The maps in the book function as argument-building devices, because they produce a spatial reading of the sources. In them, I plot the data from land grants, court testimonies, almanacs, newspapers, photographs, disputes, and recollections of the city. This additional level of analysis identifies spatial patterns that are not easily discernible from close readings alone.

CARTOUCHES

I have built a cartouche for each of the six main speculative maps for the walking cartography, developed from a collection of the sources employed to write and draw these histories. For example, the cartouches for the introduction and chapter 2 (see figure 0.8) represent the sources required to reconstruct the street network for Lagos, and to transform the land grants from social to physical network. Generally, the images in the segments of the circle follow the same pattern: They are, clockwise from top right, (R1) an image of an important protagonist or source, (R2) the Spyglass map for the chapter, and (R3) a contemporary image from the walking cartography research. The images on the left side of the cartouche, starting from the bottom, include (L3) a historical source, (L2) a historical map, and (L1) an image from the time period of the chapter. At the center is an icon of Lagos Island, and the date of the map.

Each map uses a distinct set of sources, as exemplified in their cartouches. For instance, in chapter 4, I transpose land grants (which are based on social networks) onto the city's topography, anchored by significant places like the church, cemetery, or even homes. Where in chapter 3 the data on the map are derived from a small set of land grants and a handful of disputes, the sites mentioned—the CMS mission, Government House, and so on—are on sites that are still accessible in the city today. "Ọba Dosunmu's Land Grants," the map in chapter 4, is based on a larger and richer data set—the seventy-six land grants given by Ọba Dosunmu—but because those are based on interconnected social relationships (and places that no longer exist), that map is more speculative in terms of place. In this map is a discussion of the importance of uncovering the old city, starting at its edges, then moving inward to the sites that are most meaningful.

One of the earliest mentions of Lagos comes from the seventeenth-century journal detailing Andreas Josua Ulsheimer's voyage to Africa. He indicated "Rio de Lago" on a map of the West African coast and noted its tributary relationship to the kingdom of Benin.[7] Since then, Lagos has been marked in different ways on historical maps. Even though not every map can be georeferenced, they can still be analyzed for historical data. *Imagine Lagos* employs historical maps divided into eight different primary categories based on the most important arguments they make about space. These categories are racialized, perimeter, intimate, fix-and-fill, territorial, itinerary, settlement, and missing maps.

Racialized space maps explain and contrast how Africans and Europeans inhabited space, characterized succinctly by John Glover's 1859 *Sketch of Lagos River,* especially as it resides at the Royal Geographical Society (see figure 4.1).[8] Similar to this is the *Sketch of Abẹokuta* in a missionary's report on the settlement of Lagos and the cities surrounding it.[9] It shows the different missions divided by Black and White missionaries. An important map used in chapter 5, *Lagos et ses Environs,* drawn by a Roman Catholic priest in 1876, highlights Black and European spaces in Lagos, using geometric shapes to distinguish between Eko, the "villes noire" or old town, from Lagos, the European quarter. The cartographer does not suggest a racial hierarchy but emphasize a difference in the ways space was and should be inhabited. Each map, in its own way, argues for the differences in (male) African and European spatial practice.

Perimeter maps are drawn from a distance and often represent only the edges of the city. With ready access to the city after the 1851 bombardments, fewer and fewer of these maps appear through time. Most of these kinds of maps are used in chapters 1 and 2, where the officers on the ships of the West Africa Squadron produced maps of the coastline of West Africa, and eventually three close-up versions of Lagos Island itself, such as Thomas Earl's 1851 sketch, *Lagos River.*[10] This kind of map insists on the emptiness of local spaces, casting them as sites of opportunity for European expansion.

Examples of perimeter maps include the several well-known renditions of the Bight of Benin, but the first to be used here is Robert Norris's map of the Slave Coast in the late eighteenth century.[11] Also included are the Admiralty maps of the coastline and various city views of Lagos, Porto Novo, Badagry, and others.[12] These maps are followed by the maps from the bombardment—for instance, the one by Earl (master of the HMS *Harlequin*) above. The last in this series of three maps is H. P. Ward's, drawn on the eve of the bombardment.[13] Early versions of these maps are accompanied by views of the coast, with hydrographic details for passing ships.

Intimate maps frame personal interactions in smaller spaces and are often about conflict. In Lagos, these conflicts are almost always over land. These maps and sketches are often bundled in the letters of colonial administrators and missionaries and found where those documents are archived, like in the CMS archives for the Yoruba Mission, and in the Colonial Office files of letters from consuls and governors. As struggles are often over land, these maps present these interactions at the very smallest scale. Examples include the grant map in 1855 (drawn by the request of Benjamin Campbell, the British consul) to bolster his accusations that the CMS missionaries were in effect stealing land along the shore, and the plan of Akitoye's grant to the CMS.[14] A different example, the 1871 plan of the Marina, walks the viewer along Water Street, now the old Marina, showing the owner of each plot of land.[15]

Fix-and-fill maps of Lagos often show only Lagos Island, and are filled with details of streets, infrastructure, and improvement projects. These colonial maps show a "full" city, always in need of fixing. These maps are among the most popular and can be found at the archives of the National Museum in Onikan, the British Library, Special Collections at the Cambridge University Library, and the National Archives in the UK. In these maps, we see the clearest overwriting of the original spaces of Eko. These maps serve important purposes in the book as they provide the most detailed information about the city's urban form and even offer impressive topographical information. Through these, we see that Lagos, though limited in a small space, is always evolving. These maps provide most of the visual information for chapter 1, as the street names appear most clearly on these types of maps. They also provide information on topographic elevation. Fix-and-fill maps represent Lagos at its most urban. In these maps, Lagos is constantly being improved, and in this series the city changes shape in the most subtle ways. Within these categories are smaller subsets of maps.

Imagine Lagos uses several fix-and-fill maps (e.g., see figure 0.9) from the sequence of plans of the town of Lagos, beginning in 1883 and continuing until 1942.[16] These maps best describe the changes in Lagos Island's shoreline, showing how different water features were filled in, and track the efforts to fix the drainage and other issues in the low-lying island. The best known is William T. G. Lawson's 1885 version, while the most detailed appeared as an ordnance survey in 1891.[17] This set contains fourteen large maps. Each sheet can fill a very large table. Larger still is the 1926 plan. Until

FIGURE 0.9. Detail from *Town of Lagos: West Africa (15 Sheets Mounted as One),* 1926, 1 inch to 88 feet (scale), CO 1047/676, courtesy of the National Archives, Kew, UK.

1891, the perimeter of the island was still intact. By 1898, Ẹlẹgbata Creek, on the southwestern edge, is gone, filled in because of antimalaria campaigning. Then, by 1908, so is Alakoro.[18] At least half of Idunmagbo Lagoon in the Northeast disappears by the early 1930s, and by the end of the decade, Isalẹgangan Lagoon to the east is also filled. By the 1940s, the island's perimeter is a smooth curve with little to no interruptions.[19]

Territorial maps are about colonial ambition and expansion. They visualize the expansion of British territory from the original confines of the coast. These maps show territory that has been acquired, demonstrating British mastery over space, such as the sketch of Lagos from the West Coast of Africa collection; Colonel Henry Ord's *Outline Map Showing the British Territory at Lagos* drawn to accompany his 1865 report on the West African settlements (see figure 0.10); and even Lawson's *Plan of the Town of Lagos,* which highlights in the same red color (that has now faded to pink) all the sites at that scale that are part of the colonial bureaucracy, from the police stations to the courts and the Government House.[20]

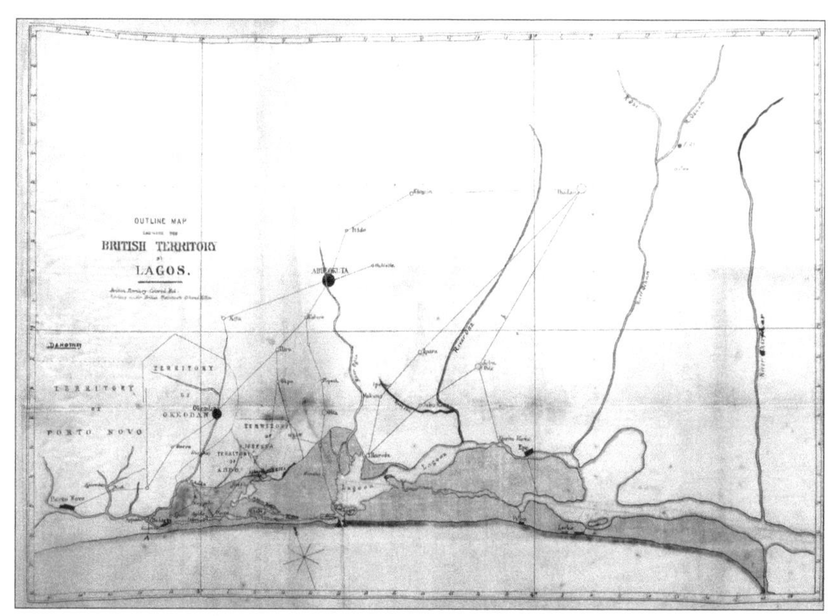

FIGURE 0.10. Detail from *Sketch of Lagos and the Adjacent Country (To Accompany Report by Col. H. St. George Ord, R.E., 1865)*, 1865, CO 700/Lagos 1, The National Archives, Kew, UK.

A unique set of the maps I read are the settlement maps (see figure 0.11). Each shows a different understanding of the range of urban and rural settlements in the Bight of Benin and reflects growing European familiarity with the interior of the Bight. Each map demonstrates the different densities of settlement patterns in the Bight of Benin. These settlements range from open cultivated lands to villages (inhabited or not) and to cities like Badagry, Lagos, Porto Novo, and Epe.[21] Despite arguments that frame West Africa as an undifferentiated, open, and mostly rural space, these maps feature geographic and settlement variety, from cities like Lagos, Abẹokuta, and Porto Novo to open grasslands, forests, and farmlands. They show how the towns were situated—mostly hugging the Atlantic or lagoon shorelines—and how rivers like the Ogun showed pathways into the interior. Interior settlements, like Ibadan and Abomey, were created according to a different geographic logic, which is reflected in these maps. Navigation was of critical importance, especially in terms of the shallow lagoons and rivers where warships were constantly grounding. Alex Tickell has pointed at the ways maps can be read as the "products of cross-cultural negotiation," some of which, in this reading, preserve moments of crisis.[22]

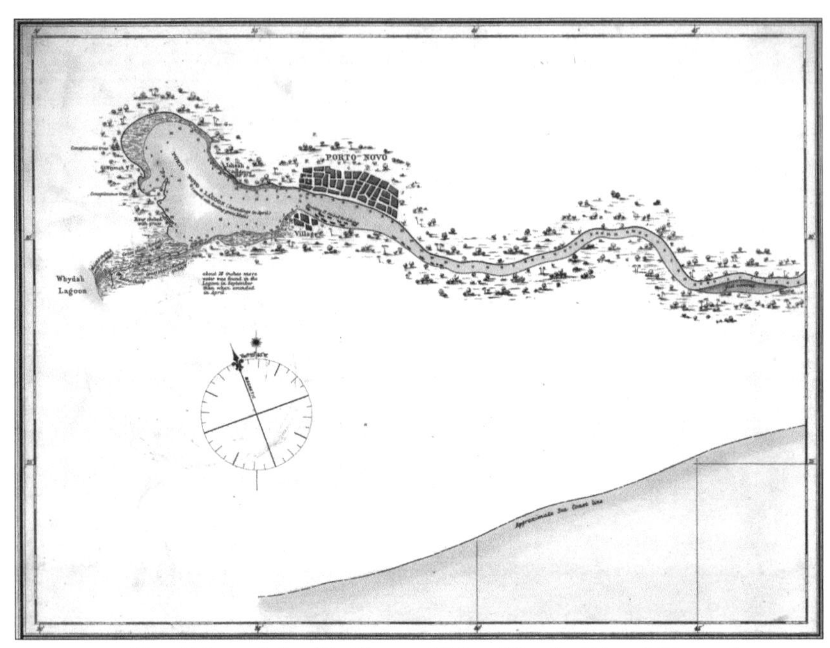

FIGURE 0.11. Detail from John H. Glover, *Bight of Benin: Inland Water Communication between Lagos, Badagry, Porto Novo and Epè*, 1858–59 and 1862, Maps SEC.11.(445a.), courtesy of the Royal Geographical Society, London.

Itinerary maps were made during expeditions into the interior—for instance, along the Ogun River. Maps such as Burton's 1892 *Chart of the Ogun River* (see figure 0.12.) offered specific knowledge about geography, terrain, and people, but also the potential for commercial exploitation. Used by explorers, these maps are often found in the archives at the Royal Geographical Society in London and are accompanied by field notes from men like Richard Burton. Itinerary maps focus on how to travel between cities, showing well-known local shortcuts like the Agboyi Creek, north of Lagos Island. Interesting examples filled with detail include the surveys of Abẹokuta in geographical context, and versions of the Ogun River in 1862 and again in 1892.[23]

Last is a set of maps that, although referred to in the documents, have not survived in the archival record. These missing maps are described well enough to help reconstruct space. Rev. Charles A. Gollmer, writing from Water Street in Lagos to Henry Venn and Major Straith of the CMS, also promised a map. "I here with send you a little map of Lagos," he wrote, "which I believe to be on the whole pretty correct."[24] When James White, a CMS missionary, wrote to Venn in 1852 to "see attached the map I have drawn for you," his letter made it into the archives, but the map never did,

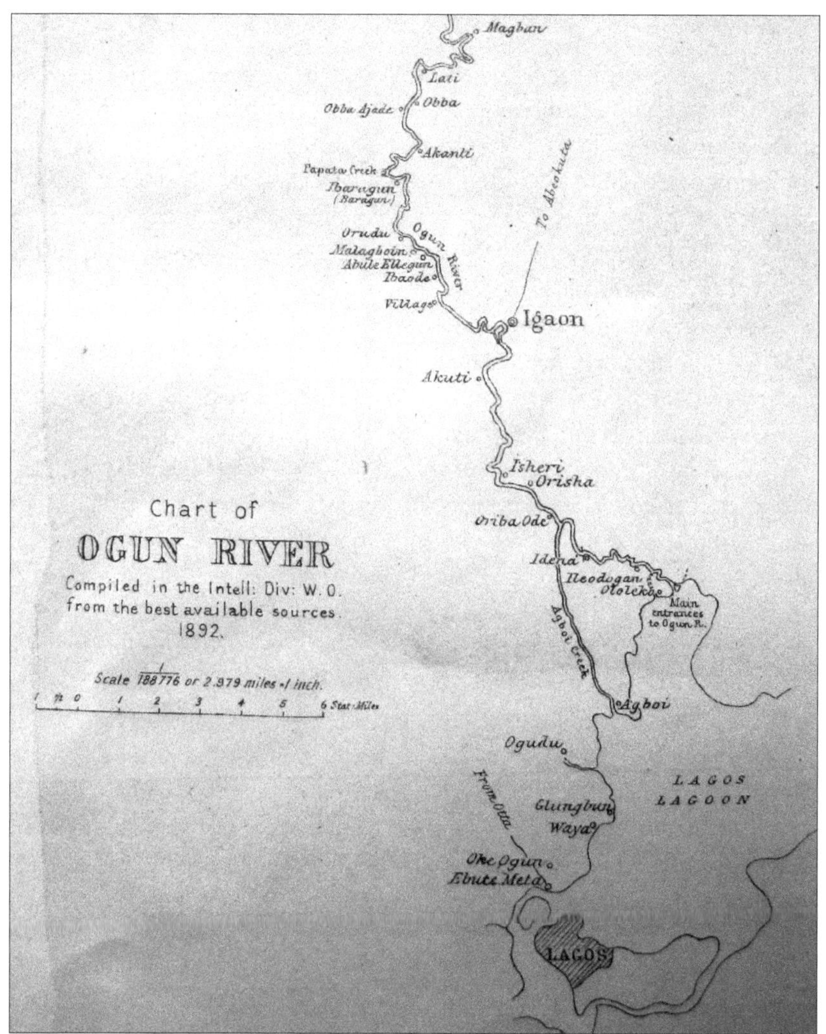

FIGURE 0.12. Detail from *Chart of Ogun River,* August 1892, MPG 1/850/4, courtesy of the National Archives, Kew, UK.

leaving the researcher to wonder sadly what this could have been about. In 1860, Captain James P. L. Davies's land grant from Ọba Dosunmu mentioned an "annual plan of Lagos," but this map has not survived.[25] Campbell, the consul, declared in 1855, "I have no Admiralty plan of the island, and I am informed there is as yet no correct one published."[26] However many maps I reproduce here, perhaps the most interesting maps are the ones that do not exist, are missing, or have been lost in time. The references to these maps are cited in chapters 3 and 4.

In writing on British maps of North America, historian S. Max Edelson points to three different ways mapmaking can conjure up the idea of a place: it represents "what those places were, how they came to be, and what they should become."[27] A contemporary map of Lagos reveals layers of the past, if you know where to look and have the tools to do so. These maps obscure the missing features that have been papered over or filled in the name of slum clearance, sanitation control, and other "improvements" that were part of the colonial civilizing mission. Since 1861, the island's perimeter and interior have been remade through dramatic human-made and natural shifts and expansions, leaving it unrecognizable in its contemporary form when compared to the nineteenth century.

Since the late nineteenth century, new regimes have filled in, expanded, and altered Lagos Island's outline. In the interior, administrative neglect was followed by slum clearance schemes that disrupted people and place in quarters like Ereko. However, it is possible to discern the mid-nineteenth-century outline of the city, which provides important spatial context for places that still exist, and those that have been written over in time.

From above, the shape of old Lagos is easy to discern. In the northwest, it is visible in the curve of Adeniji Adele around Isalẹ Eko, in the curve of streets like Ọdunfa and Evans in Isalẹgangan, and in the gap created by the MacGregor Canal in the middle of Lagos Island. It is less obvious on the southern edge of the island, where the expansion is parallel to Water Street (now Marina Road).

The first speculative map (see map 0.4) is one of Lagos Island that reveals the historical shoreline of the city and the topographic information about the island's geography. This map shows the original outline of Lagos Island in the mid-nineteenth century. This old perimeter of the island is etched in black on top of a satellite map of contemporary Lagos, showing the difference in the historical boundaries of the city in a contemporary context. The expansion of the city gave several sites historical importance, such as Ẹlẹgbata Creek and Alakoro Swamp and Island in the western edge of the island. MacGregor Canal, cut in 1899 to help drainage in the city, had divided Lagos Island into two, separating the habitable portions from the parts still covered in swamps and mangroves.

The cartouche for this map contains six images that summarize the sources used to create this map: (clockwise from top right) a drawing of Rev. Samuel Ajayi Crowther with the queen and prince of England in 1851; a Spyglass over the original site of Isalẹgangan Lagoon; a 2018 photograph of the remaining mangroves in Lagos; a drawing of enslaved people being shipped from the beach in Lagos; Ẹlẹgbata Creek from the 1891 *Plan of the*

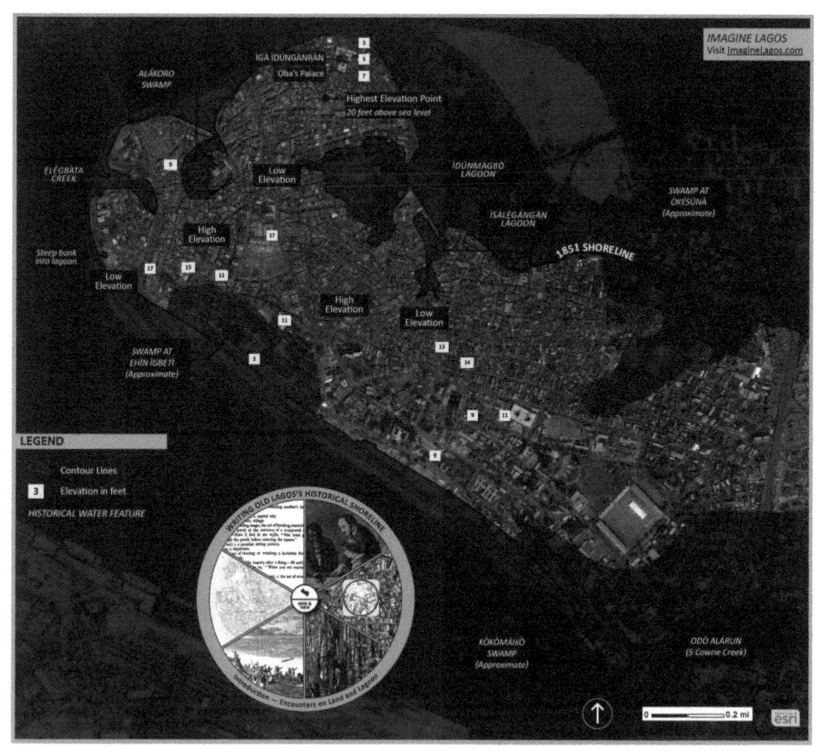

MAP 0.4. "Writing Old Lagos's Historical Shoreline." Base map courtesy of ESRI. Images in the cartouche are, clockwise from top right, as follows: (R1) Rev. Samuel Ajayi Crowther showing a map of Lagos to Prince Albert, Queen Victoria, and Lord Wriothesley Russell at Windsor Castle in November 18, 1851, in Herbert Samuel Heclas Macaulay, *Justitia Fiat: The Moral Obligation of the British Government to the House of King Docemo of Lagos; An Open Letter* (London: Printed by St. Clements Press, 1921); (R2) Spyglass over the former site of Isalẹgangan Lagoon, courtesy of ESRI; (R3) mangroves around Lagos Island, photograph by author; (L3) "Shipping Slaves through the Surf, West-African Coast: A Cruiser Signalled in Sight," 1845, in "Slave-Trade Operations," *Church Missionary Intelligencer: A Monthly Journal of Missionary Information* 7 (November 1856): 241–55; (L2) detail from *Plan of the Town of Lagos—West Africa (in 15 Sheets) (Survey in Fourteen Sheets Numbered 1 to 13 and "11 and 15" and Bound as a Volume),* 1891, 1 inch to 88 feet (scale), 63 × 90 cm, CO 700/Lagos 14, The National Archives, Kew, UK; (L1) detail from Samuel Crowther, *A Vocabulary of the Yoruba Language* [. . .] (London: Seeleys, Fleet Street, 1852). Map by author.

Town of Lagos; and a page from Crowther's Yoruba-to-English dictionary, highlighting Yoruba conceptions of space.

The outline itself is based on sixteen georeferenced perimeter maps and two perimeter sketches read for their spatial content. The earliest maps of the entire island that aspire to geographical accuracy are the fourteen maps in the 1891 ordinance survey.[28] Charting this outline allows us to visualize the location of the old city, relative to contemporary Lagos. The topographical data, especially the elevations, come from the plans of Lagos drawn in 1904, 1911, and 1926.[29] The analytical maps represent the contents of a chapter's argument, are a visual representation of the argument, or suggest a specific reading of space, supporting the argument in the chapter.

The Marina today may appear like the edge of the island, giving the impression that there was a wider buffer between buildings that were originally the CMS compound and Government House. (See, for instance, map 5.2.) In fact, the original southern edge of the city was farther inland. In the northeast, the only traces of lagoons, like Idunmagbo and Isalẹgangan, are in the shapes of the streets that echo their presence. The same is true in Alakoro and Ẹlẹgbata in the west.

Each new map presents a set of recovered, discovered, or interpreted features (such as a slave barracoon or cemetery) that gain importance in the period covered in the chapter. The data comes from maps as recent as the 1904, 1911, and 1926 plans of Lagos[30] as historical clues continue to accumulate into the mid-twentieth century.[31]

For instance, the process to ascertain the location of pre-bombardment Lagosians in chapter 1 requires the comparison of at least three historical maps with a court case from the late nineteenth century. The compound of Ṣẹtẹlu, alias Gomez, is important for a variety of reasons, especially because there is so little documentation of ordinary folks before and after the bombardment; his is one home that seems to have survived the bombardment in 1851, establishing him as a longtime resident of the city. In this case, his home can be located with a certain degree of accuracy because, by 1877, the John Holt company had acquired his land. We know this through a lawsuit it filed against the government to determine its rights to the foreshore (sloping land to the coast) in front of its warehouse on the Marina. During the trial, it emerged from the testimony that the company bought the land from a commercial firm, who had purchased it from Ṣẹtẹlu. However, Holt's premises do not appear on maps until 1904. In another example, the place that was said to have been a slave barracoon in the 1840s (see chapter 1) became a church in 1851 (see chapter 2), and is now the site of St. Paul's, Breadfruit. See map 0.5 for how locations have changed in Lagos.

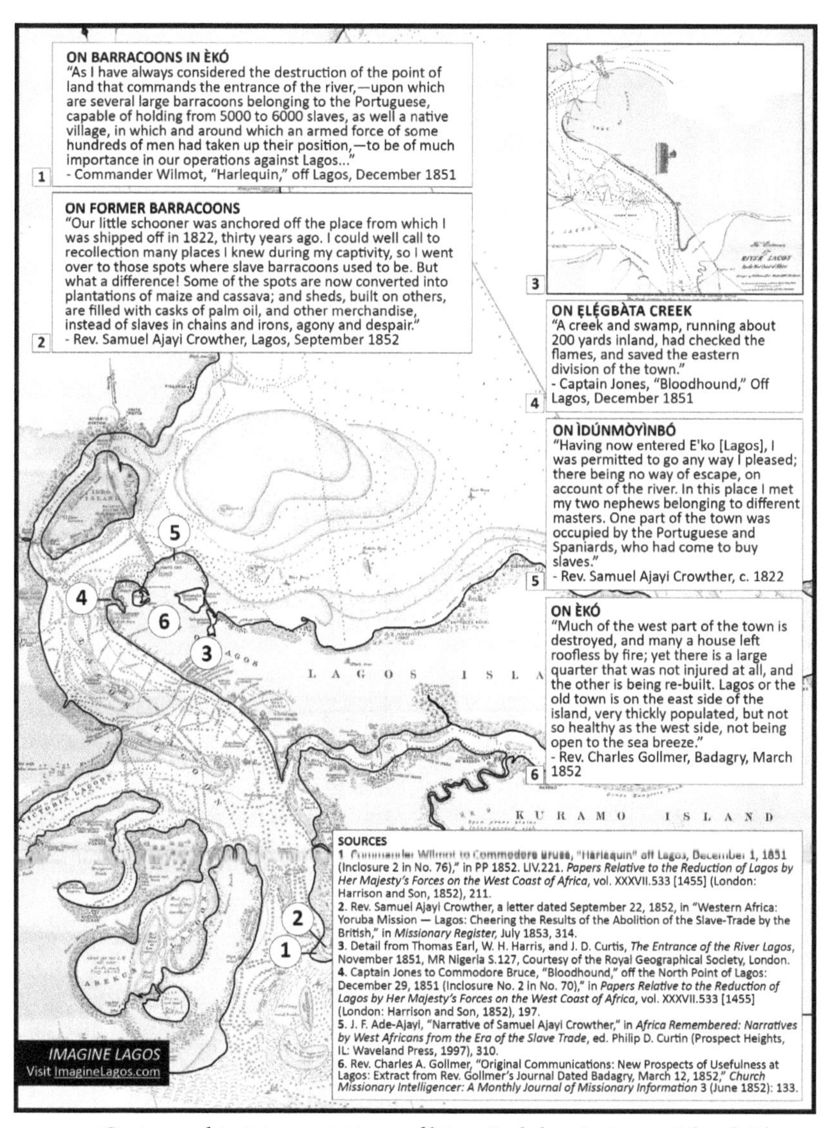

MAP 0.5. Cartographic interpretations of historical data in Lagos Island. The historical map is a detail from William C. Speeding, *Lagos Harbour,* April 1898, 1 inch to 1,600 feet (scale), 63 × 90 cm, and is courtesy of the National Archives, Kew, UK. The numbers on the map represent the sites that are the visual interpretation of the historical data. Map by author.

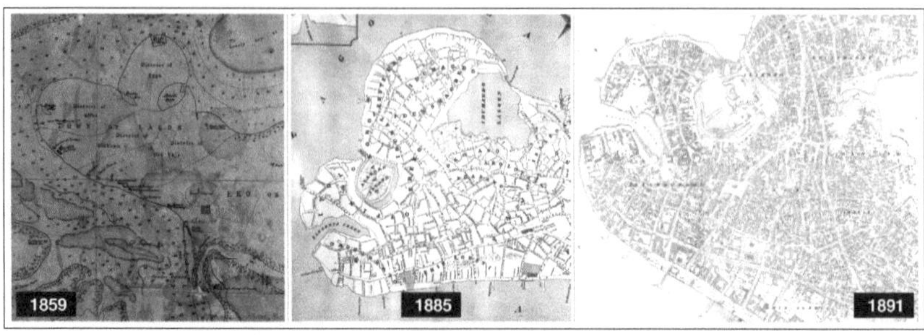

FIGURE 0.13. Comparing representations of residential quarters in Lagos. *Left to right:* detail from John H. Glover, *Sketch of Lagos River,* 1859 [corrected to 1889], 1:20 scale, MR Nigeria S.127–24, courtesy of the Royal Geographical Society, London; detail from W. T. G. Lawson, *Plan of the Town of Lagos, West Coast of Africa: Prepared for Lagos Executive Commissioners of the Colonial and Indian Exhibition 1886; By Order of Fred. Evans, Esq. CMG. Deputy Governor,* 1885, about 1 inch to 275 feet (scale), CO 700/Lagos9/1, courtesy of the National Archives, Kew, UK; detail from *Plan of the Town of Lagos—West Africa (in 15 Sheets),* 1891, 1 inch to 88 feet (scale), 63 × 90 cm, CO 700/Lagos 14, courtesy of the National Archives, Kew, UK.

MAPPING RESIDENTIAL QUARTERS IN OLD LAGOS

Figure 0.13 shows the speculative extent of residential quarters in old Lagos in 1851, based on the comparisons of three fix-and-fill maps from the nineteenth century. The city had grown from the six original residential quarters—Iga (a.k.a. Isalẹ Eko), Ọfin, Olowogbowo, Ereko, Itọlọ, and Oko Faji—to a documented twenty by the middle of 1872.

Since the 1850s, cartographers have been trying to figure out the extent, borders, and centers of the residential quarters. Historians have gestured statically toward these spaces without regard for their spatial contexts. Evidence shows that people who walked in Lagos were very conscious of the different districts they were in. CMS missionaries like James White and Charles Gollmer were careful to distinguish between the different sections of the city where they lived and proselytized, and where they eventually built their churches and schools, like Idunmọta and Ebute Ero. Where do the quarters begin and end? Why do they have particular names? How are these names related to the *idẹjọ* chiefs, who were known as the land-owning chiefs in Lagos? What distinguished one quarter from another?

Figure 0.13 shows three nineteenth-century maps of Lagos that attempt to outline the historical quarters. The map on the left is a section of Glover's 1859 *Sketch of Lagos River,* which is the oldest visual representation of the

MAP 0.6. Investigating residential quarters in Lagos in 1851. The suggested boundaries of the quarters come from the 1891 survey of Lagos Island: *Plan of the Town of Lagos—West Africa (in 15 Sheets),* 1891, 1 inch to 88 feet (scale), 63 × 90 cm, CO 700/Lagos 14, The National Archives, Kew, UK. Map by author.

residential quarters in Lagos. However, he only shows four of the quarters and calls Isalẹ Eko the "Iga quarter."

In the center is Lawson's 1885 *Plan of the Town of Lagos.* This is one of many fix-and-fill maps that represents the different quarters with a label, but there is no attempt to show the boundaries, or explain how each quarter is distinct from the others. The map on the right is a georeferenced combination of four sheets from the 1891 *Plan of the Town of Lagos.* This is the first map in which we see some articulation of the different quarters in the city, based on the streets and intersections on the island. Even though this map was created two decades after the period covered in the book, a careful comparison with the text and cartographic record suggests some of the boundaries outlined in the map and the ones that follow (such as the 1904 *Plan of the Town of Lagos*) has some resonance with the historical understandings of space, place, and centers in the city.

Map 0.6 (which contains the quarters recorded by 1852) shows what happens when one extracts these quarters from the maps themselves and compares them to each other. The quarters' shapes are all drawn to scale, but arranged outside of the historical shoreline so they can be visually compared and analyzed for patterns.

In map 0.7, we see the rapid expansion or representation of Lagos from six quarters to nearly two dozen by 1872. These data are important for interpreting the material on Ọba Dosunmu's land grants (explored in chapter 4) and in thinking about the significance of different streets and social landscapes in Lagos. Looking at these quarters individually is important, especially for drawing attention to historically neglected quarters like Idunmọta, Arọlọya, and Idunṣagbe.

In *Isale Eko Day,* the Descendants' Union describes Isalẹ Eko as bound by Idunmagbo Avenue, Nnamdi Azikiwe Road, and Adeniji Adele Road, a description that is a close match to the historical quarters noted as Iga on maps of the 1850s.[32] But these references change by the 1870s. Three

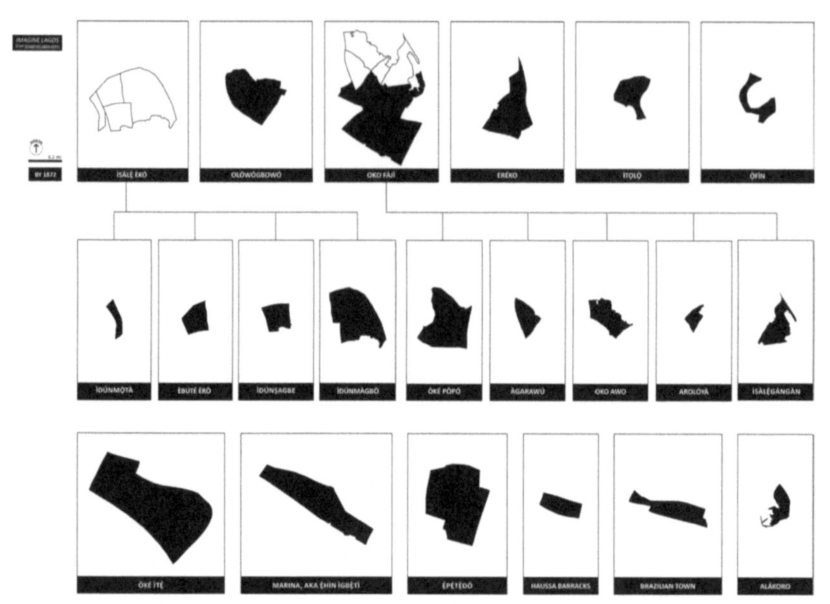

MAP 0.7. Investigating residential quarters in Lagos in 1872. The suggested boundaries of the quarters come from the 1891 survey of Lagos Island: *Plan of the Town of Lagos—West Africa (in 15 Sheets),* 1891, 1 inch to 88 feet (scale), 63 × 90 cm, CO 700/Lagos 14, The National Archives, Kew, UK. Map by author.

particularly important patterns emerge when studying the spatial material of these quarters. The most interesting difference is how the Iga, or Isalẹ Eko, quarter is split into four smaller quarters in the maps of Lagos (Idunmọta, Idunṣagbe, Ebute Ero, and Idunmagbo); and how the area historically described as Oko Faji, or the fields at Oko Faji, yields new quarters, including Oko Awo, Aroloya, Agarawu, Isalẹgangan, and Oke Popo. These new quarters emerge in the grants distributed after the bombardment, and are defined by the streets that possibly form their boundaries and the landmarks that form their core, such as markets, open squares, and the *iga* (palace or seat) of important chiefs and merchants. By 1891, the districts are marked out clearly on maps, even though the shapes are difficult to discern without referring to the cartographic material found in the ordnance surveys.[33]

1 ~ Streets, Placemaking, and History in Old Lagos

At Omididun and Inabẹrẹ

> Massey Street? Keep going. Ah, I know the place. Turn left at the junction. Keep going. You'll see a green building, turn right. You should burst out on the right road. If you see a vulcanizer, you've gone too far. When you see Aunty's shop, ask somebody.
>
> —Directions given to me for Massey Square on Lagos Island, July 2018

TODAY, ASKING for directions on Lagos Island can send you on a tangled trip. The directions include mobile and fixed landmarks, sudden turns and convoluted routes, and the occasional street name, and the trip often requires you to speak to more than one person. I would later discover that directions to get around in mid-nineteenth-century Lagos were not much different. Starting in 2018, Pẹlu Awofẹsọ,[1] a renowned travel writer; Olalekan Adedeji, a photographer; and I were searching for traces of an old set of railway tracks for a digital mapping project on the history of the early twentieth-century steam tram.[2] The old tram's main line had carried Lagosians across the old Marina, all the way from the Colonial Hospital to the King's warehouse, before heading north past Ebute Ero to join the government railway at Ido. The tram's secondary line had run through the center of Lagos Island along Tokunbọ Street (which marked the northern edge of

MAP 1.1. Spyglass over the junction of Omididun Street and Inabẹrẹ Street in Lagos Island. Base map courtesy of ESRI. The historical map is a detail from *Plan of the Town of Lagos—West Africa (in 15 Sheets),* 1891, 1 inch to 88 feet (scale), 63 × 90 cm, CO 700/Lagos 14, courtesy of the National Archives, Kew, UK. Map by author.

the Brazilian quarter, which was demarcated in 1852 and bounded roughly by Igboṣere Street to the south, Tokunbọ Street to the north, and Cow Lane to the west). As we walked around searching for any trace of the tracks or the tram station, we found ourselves at the intersection of Glover and Tokunbọ Streets. This intersection felt like an auspicious site, because of colonial administrator John Glover's influence on the settlement of immigrants from Brazil, his plans for constructing early colonial Lagos as a free space, and the settling of Oṣodi Tapa's descendants and dependents in the new Ẹpẹtẹdo quarter in the 1860s. Further research into the development of streets, new quarters, and justice in Lagos (as seen in chapter 5) would reveal insights about Glover's relationships with Lagosians.

Within a few minutes of walking to the west, we reached Massey Square, the junction where Tokunbọ Street becomes Massey Street. We came upon the most unusual names of streets. Omididun Street runs north from Tokunbọ Street. In Yoruba, *omididun* means "very sweet or tasty water." North of Tokunbọ, Omididun Street intersects Inabẹrẹ Street, another very striking name. *Inabẹrẹ* means "the fire began or started here." Fires in colonial Lagos were frequent, expected, and quick to spread. They also left little trace in their wake. So finding this street, its name a street-archival trace of those fires, was particularly compelling. On the one hand, the street names draw our attention to, on the one hand, the ordinary peoples whose lives were shaped by the lagoons, creeks, and swamps of Lagos and so often tragically affected by these conflagrations and, on the other, the value of the symbolism invested in the colonial project of naming the hitherto unmarked spaces of Eko.

Walking through the city now, and reading the street names, clarifies that evidence for the city's past exceeds what we can find in documents alone, that the social landscape of street names, quarters, squares, and neighborhoods is itself an archive of the city's past. For inṣtance, looking west from Ita Balogun to the site of the old cemetery (at Joseph Street and Campbell Street), or south from the Iga Idunganran (the palace of the ọba, or king, in Isalẹ Eko), one sees spaces which contain a network of streets, lanes, and avenues that narrate the city's history through the symbolic value of names. What kinds of stories does this network tell?

This chapter spans the period of old Lagos—that is, the city as it evolved between 1845 and 1872. It focuses on a space made possible by the interventions and influence of three powerful men: Ọba Kosọkọ, his right-hand man, Oṣodi Tapa, and Glover, the British naval officer turned cartographer and administrator who filled a variety of administrative roles from 1859 to 1872 in the Lagos area. He was appointed lieutenant governor in 1863 and functioned as administrator starting in 1866. He left Lagos colony for good in June 1872.[3]

However, this chapter does not dwell on elite figures, but instead focuses on the histories of people and places made possible by their influence in Lagos. What do the names of streets have to tell us about the histories of old Lagos? It is not a history only of ọbas or of chiefs, or of British imperial rule. The history embedded in the street names points to narratives of loss, the pain of enslavement, the prominence of a new religion, while pointing also to the tenacity of worship of *oriṣa* (Yoruba deities) around Eko. The traumas of slavery—both domestic and transatlantic—and the possibilities around gender relations and religious worship that predate British colonialism all emerge from the names of streets across the island.

Prior to the 1870s, there was no publicly recorded list of street, square, and compound names in use in Lagos. So how did Lagosians navigate their city and island from the 1840s to the 1870s? Until the 1850s, the way space was marked was not legible to the visitors who compiled the letters, sketches, and maps of the town, and there was also no written public system of landmarks, homes, compounds, or markets. Yet these sites held specific significance for those who lived, worked, escaped from, or were even enslaved there. These significances, I argue, are archived in the symbolic value of street names.

In February 1868, John Payne, an English-speaking Yoruba court clerk, named all the newly paved streets, lanes, and roads in the city of Lagos

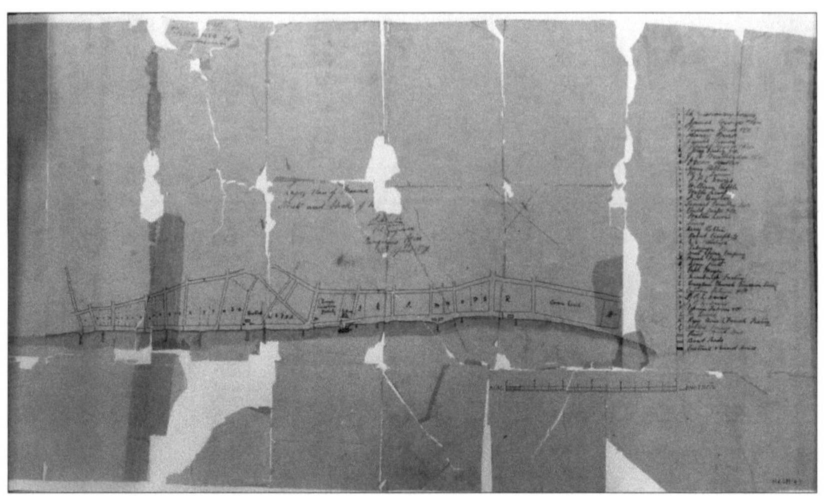

FIGURE 1.1. Detail from J. Hunter, *West Africa: Lagos (Now in Nigeria); Plan of the Marina Area, Showing Streets, Blocks of Buildings and the Owners of Property,* April 13, 1871, 1 inch to 66 feet (scale), MR 1/389/4, courtesy of the National Archives, Kew, UK.

in the wake of a new colonial infrastructure that was quickly taking root.[4] Rather than superimposing British names or even a stripped-down numbering scheme, he chose mostly indigenous Yoruba names that reflect a remarkable archive of local priorities, symbols, events, places, and people. These names highlighted spaces already in use (like Isalẹ Eko for the core of the old city), formerly enslaved men who had returned to Lagos (such as Tokunbọ, Joseph, and Bishop), and even sites where major fires began (Inabẹrẹ). This chapter uses the first written network of street names in Lagos Island to enter into the spatial history of the city in the mid-nineteenth century.

These names were chosen in the first years of formal British rule in Lagos, a period when the priorities and perspectives of the British framed written, visual, and cartographic representations of the city. In naming the places in this colonial city after indigenous priorities, however, Payne was asserting a particular identity for the city that retained specifically curated aspects of its past and pointed toward future possibilities as a colonized space. The names he chose suggest the city Payne marked in place was Yoruba in culture, male in gender, Christian by design, and tuned to the legitimate commerce of cotton and indigo and not slavery. But despite these larger ambitions, he also fixed competing claims; several of the men who were now wealthy and/or Christian had been formerly enslaved, and foregrounding Yoruba names meant that names of orișa and water bodies crept into the scheme. Payne fixed these competing claims against the claims of a colonial imaginary that threatened to overdetermine the city's symbolic value by shaping Lagos from a nonindigenous perspective. (See figure 1.1. for a view of the list of streets intersecting Water Street.)

In 1966, James Baldwin noted that "history does not refer merely, or even principally, just to the past." What gives history its "great force," he continued, is "how we are unconsciously controlled by it in many ways [as it] is literally *present* in all that we do."[5] In fixing these particular histories of the city in place in 1868, Payne produced an archive of Lagos's past. When he named streets like Apọngbọn in the Itọlọ quarter, Bridge Street in the Idunmọta quarter, Ọbun Eko in the Ebute Ero quarter, and Oluși in the Isalẹgangan quarter, he was marking space and drawing attention to narratives around power, gender, religion, and freedom in midcentury Lagos. Just as court transcripts and letters can be read and analyzed for insights into the past, streets in Lagos bear witness to the ways the city's pasts intertwine. At the junction of some of these pathways are the stories of ordinary Lagosians who flit in and out of the archival record.

The lanes, avenues, and roads were not inventions out of whole cloth (as they were based on existing roads), nor do they reflect a totally indigenous version of town planning. Rather, I suggest that, in marking out spaces, connecting places, and embedding intentionality in terms of direction, every street that appears on a map, grant, or even newspaper article narrates a set of possibilities precisely when Eko was being transformed into a colonial city: Lagos. My research shows that these street names provide a transcript of the history of Lagos, arranged around the questions and interests of locally generated debates in the city. By mapping and analyzing the layers of these streets through space and time using historical GIS, this chapter reveals how crises over enslavement, colonialism, and religion were debated and marked in space.

Today, greater metropolitan Lagos is a space where history as a subject is routinely removed from the curriculum of local schools, and fly-by-night demolition schemes of landmarks frequently erase the architectural legacy of the island.[6] In this chapter, I use land grant records, maps, photographs, and texts that documented the city in the nineteenth and twentieth centuries to transform the street names from a patchwork of nouns, names, and metaphors into a legible and engaging transcript of Lagos Island and its history. Streets and places thus emerge as sites of historical memories, and my own maps of the mid-nineteenth-century city set priorities for analyzing

FIGURE 1.2. "Lagos, WestKüste von Africa," 1859. Reproduced by kind permission of the Syndics of Cambridge University Library.

and understanding the histories of encounters and spatial drama in Lagos, which emerges here as a single space, rather than a sum of its quarters. A focus on the city as a whole—as both physical and social landscape—brings us closer than ever to understanding the lives of Lagosians. Between 1845 and 1872, in old Lagos, the streets keep these histories. Why? Because street names in Lagos are both typical and unusual.

LANDSCAPE AND GEOGRAPHY AS ARCHIVE

The street names in mid-nineteenth-century Lagos cluster and spread, forming a variety of interesting patterns over the island. Of those patterns that are legible, two are particularly interesting: the pattern of names connected to orișa worship, and those named after the formerly enslaved.

Tracking historic street names reveals a new geography of Lagos Island as the streets stretch outwards from the older settlements in the west in Olowogbowo, Tinubu Square, and Balogun. Streets and their names provide a stable anchor for the people, who are almost like fugitives in the archives of their own city, Eko. The nineteenth-century city that emerges from these street names is an uneven collage of lanes, roads, and broad thoroughfares, intersecting to form tight spaces or open ones, which Europeans imagined as "squares," such as Tinubu Square (Ita Tinubu) in Oko Faji or Balogun Square (Ita Balogun) between Olowogbowo and Ereko. However, European narratives and maps rarely venture north of Tinubu Square into the "Old Town" quarters, such as Agarawu, Ọfin, and Aroloya. But these are the new quarters that the street names stretch into. There is no doubt that the indigenous Lagosians had their own names for areas, and the evidence points to the fact that the streets were named after influential Lagosians, for instance, Ita Tinubu and Ita Balogun. The indigenous names of the quarters were also in use, from Olowogbowo to Faji, and Idumọta and Isalẹgangan.

The most striking thing about the twentieth-century maps is how the nineteenth-century city is embedded within the extended twentieth-century shoreline. Lagos has expanded and been filled in. The swamps and large lagoons are gone, and only the shapes of the streets there hint at the former geography. The shoreline is split, and the island has been sliced in half by the MacGregor Canal. The earliest detailed visual transcript of the city's streets comes from William T. G. Lawson's 1885 map of the city, which he produced for the Colonial and Indian Exhibition in London.

There are also cues to physical sites that have disappeared due to the reinterpretation of the shoreline: Lake Street reminds us that the Ẹlẹgbata Creek used to dominate the west end of the island, dividing that edge between Olowogbowo and Ọfin. And, of course, in December 1851, the lake saved that portion of the town from being destroyed.[7] The streets in Lagos are often based on patterns that echo a physical past: they curve around former lagoons, point straight to public squares, and distract you from forbidden or dangerous areas. Until the 1850s, the inhabited parts of the city were largely confined to two islands—Eko and Ido—and it was not until the late 1860s that Ebute Mẹta was planned out.

The nineteenth-century sources provide some sense of the orientation of the city and its quarters. In his research on streets in Lagos, Adeoye Deniga differentiated between Ẹhin Ọgba (later known as the lower part of Victoria Road) and Ẹhin Igbẹti. The former as "behind the neighborhood" while the latter is loosely translated as "behind the wall." On the fix-and-fill maps of Lagos from 1883, 1885, and 1891, Victoria Road is not part of the Ẹhin Igbẹti area, but is immediately north of Ita Tinubu.[8] Spatially, it could represent the southern and western limit of the inhabited parts of Lagos before 1851. Deniga described it as "a jungle then, seldom traversed."[9]

We know that, historically, Lagos had distinct quarters, but we do not know what their cores were, or what distinguished one from another. Where, for instance, did Olowogbowo end and Ereko begin? Were the boundaries cultural or spiritual, or were they actually ever physically marked? The earliest attempts to map these perimeters came with Glover's 1859 sketch, but they were not connected to the streets themselves until the ordinance survey of 1891. At the core of most indigenous districts was a place of some significance, whether it was the palace in Isalẹ Eko, or even Faji's shrine in Oko Faji. In places like Olowogbowo, these significant places were likely wherever Oṣodi Tapa and his Brazilian and Portuguese allies made their bases.

The boundaries of the different quarters in Lagos are an important issue with clear cultural and personal significance, but they have been challenging to inscribe in space and effect on paper. An 1891 ordinance survey made the most complete attempt and offers important clues to the ways the edges of these districts were determined. If we use these lines as a guide to what the original lines could have been, we see that there is resonance with the sources from the 1840s and 1850s. Of course, without contemporary evidence from the time, these shapes are estimates at best, though they allow us to try to comprehend the city from above.

Glover's 1859 perimeter map is an interesting example of interpreting the boundaries of old Lagos. In marking out the original quarters, he drew lines inward from the southern coast to indicate the boundaries between them. However, his use of dotted lines suggests something about the permeability of these boundaries, as if his lines were just suggestions, or pointed to something porous or unfixed about the barriers. In his maps, Ọfin (labeled "Offee" on the map) and Oko Faji are the largest quarters, but in more accurate maps, Ọfin is one of the smallest districts in the western portion of the island.

He did not mark Ẹhin Igbẹti separately, but the buildings that he identified, from the Lagos warehouse to the French factory, brick works and powder magazine, were all of European origin. The only indigenous structure he pointed out was the king's palace, in the northwest corner of the island, and even that was made of European bricks. He also marks out three landing places, or wharves—known in Yoruba as "Ebute."[10] However, rather than use the names for the water bodies next to these landing places, he called them by the residential districts they were part of: Ebute Ero, Ebute Ọfin, and Ebute Iga.[11] On the mainland, he marked out Ebute Mẹta correctly, and even translated it to English.

Glover's map makes it clear that there are specific boundaries between the quarters, but he could articulate these boundaries only along the southern edge of the island. His map contains information that seems confusing at first. Contemporary understandings of these boundaries evolved and changed over time. The historical evidence becomes firm in 1891, with the ordinance survey of 1890–91, which used streets and roads to mark firm boundaries between the various quarters.[12] Before this, mapmakers usually suggested the extent of residential areas by simply writing the names of districts on maps.

Some areas form discernible shapes from above, and the geography of these neighborhoods is clear from the nineteenth-century transcripts. The most striking is Ereko, whose west-leaning triangle seems to point to the mainland, and is bordered by Victorian and Balogun Streets, with Broad Street forming its base. The Ẹpẹtẹdo area is nearly a square, with five north–south streets running through it. Ọṣodi Tapa was known to a be a close friend of Governor Glover, so it is fitting that the area known by that name is bounded by Glover Street on the west. Today, in addition to Ọṣodi Street, the Ẹpẹtẹdo area is run through with other streets bearing the names of subsequent governors and administrators, such as, moving west to east, Freeman, Dumaresq, and Patey.[13] The eastern border has a Muslim cemetery.[14] From the historical maps of Lagos, it seems Ereko should be the heart or center of the old city, but its associations with Kosọkọ make this a fraught claim. At Ereko's center is the Ereko Market, where Akitoye made Kosọkọ the olọja of Eko in the 1840s. Geographically, the district is adjacent to every other important one: Isalẹ Eko to the north, Olowogbowo to the west, Ẹhin Igbẹti to the south, and Ita Faji to the east.

There is a constellation of streets around the palace in the Iga quarter. The area around the palace contains the oldest settlements of Lagos Island, and the names of streets that surround the palace hark back to their origins and speak of the important indigenous inhabitants of the city. Street names in use around the palace include Ẹhin Iga, Onilegbale, Upper King, Iga Okuta, Ita Ọmọ (which leads west and branches off to Kutere Street, Idi Ọmọ Street, and Egbe Street).[15] Other streets connect to the interior. Historians have pointed to the Benin linguistic roots of Idun, a popular prefix for sites and streets in Lagos,[16] as clear evidence of the significance of Benin to Eko's origins, seen in quarter names like Idunmọta (the base of the Ọta people) and Idunṣagbe, which date to before 1850, and the names Idunmagbo Lagoon and Iduooyinbo. That many of these names are in use in the Isalẹ Eko around the palace suggests that the much-debated influence of the Bini Kingdom in the origins of Eko's ruling families was significant. Histories of Lagos have not always been kind to Ọba Akitoye, despite the pivotal role he played in the abolition of slavery in the city. There was no Akitoye Street in mid-nineteenth-century Lagos Island, even though Kosọkọ and Ọṣodi both had streets dedicated to them. By the early 1870s, the street names were more widely in use in letters, documents, and publications. Even Ọba Dosunmu eventually addressed his letters from King Street.[17]

In Lagos, like in many other settlements nearby on the lagoons, the ara Eko—inhabitants of the city, though not necessarily indigenous ones—lived in compounds, or *agbo'le,* made up of groups of houses built around a courtyard. Typically, there were one or two entrances that opened up to paths that linked different compounds. These compounds together would then form a quarter, with the streets converging at the home of the leader or chief and a market, which was the focal point of this arrangement.[18] Scholars have argued that in this scenario "the compound and not the road was the symbol of articulation."[19]

"Streets," writes Allan Jacobs, "moderate the form and structure and comfort of urban communities."[20] Streets and roads are sites of encounter, and whether these meetings are political, economic, or even religious, they are always public. Most of the sources on Lagos observed people n'ita—that is, outside, in the streets, thoroughfares, and squares that joined one street to another and fit between quarters. These streets intersected frequently and seemed (at least to outsiders) to be made of the leftover spaces between homes, compounds, and markets. There were, however, important homes (of chiefs and the ọba), large periodic markets (like Faji and Ereko), and open squares (like Ita Balogun and Ita Faji) where significant public functions were held.

By the late 1840s, Lagos's streets were noted for their narrowness and the way they frequently intersected each other. The utility of the narrowness of Eko's streets became apparent when the city was under attack, making it clear to attackers that it was hard to win a war with combatants who were so easily concealed and had home field advantage. These defensive arrangements were made so that almost every compound or house could be armed, as if it were a small fortification. Subject to destruction due to frequent fires as well as the aggression of regional powers, houses and other structures were often made of easily reconstructed and locally sourced materials. In Eko, people built their houses with lagoon mud, with walls approximately eighteen inches thick and varying in height from three to nine feet. Most of the eastern part of the island was swampy and not effectively reclaimed until the twentieth century. Some perimeter maps, including Glover's *Sketch of Lagos River,* John Pagan's *Sketch Plan of the Town and Island of Lagos,* and William C. Speeding's *Lagos Harbour,* show the settlements found on Eko and on Kuramo Islands in the nineteenth century.[21]

Some street names reflected the original nature of the spaces defined by the road. While Wiwọ Onitẹrẹ in Olowogbowo now appears quite straight on the maps, its name translates to "crooked alley."[22] Oju Iyari, a street mostly documented from 1869 to 1870, is not on the maps now, but most likely was in honor of an earlier ọba.[23] Some streets are at the center of the new districts that bear the same name, as with Oke Popo. These street names show a city with people of varied origins. Other important streets in the area include Dosunmu Street, Ebute Ero, and Ẹnu Ọwa. Deniga called Ẹnu Ọwa the "principal square" in old Lagos and an important place for "mass meetings." It is also the location for the capping of new obas in Lagos. He added that "the King's entrance" was an "open space at the intersection of upper King and Great Bridge" Streets where people paid tribute to the ọba of Benin.

Lagosians live with this history every day, but the pressures of life in the city obscure these historical echoes. The city bears, and bares, evidence of its own past. First, it carries the names of places and protagonists, both forgotten and treasured, in its street network. Its shape also retains clues to lost and almost forgotten events and reconstructions. In this chapter, I focus on one type of reclamation: recovering the names of people, events, and places inscribed into the street network of 1868. Creating a list of these names offers clues to the histories that were worth focusing on at that point. What was erased, and what remained, is an important list of local priorities and ambitions.

Some of Lagos's street names show how the island, and eventually the city, is connected to communities across the lagoons, like Badagry, Ẹpẹ, Ikorodu, and even Benin. The island was home to many important markets, of which Ọbun Eko, in the northwest, was one of the most important. People came by canoe from Badagry and Ikorodu, and went as far as Ijebu to buy cattle, food, and other items. The name Ọbun itself is said to be of Ijẹbu origin and means "market." Following that logic, Ọbun Eko would simply be "the Lagos market."[24]

Figures 0.7 and 4.1 show the island was ringed by a "dense mangrove bush," and these mangroves still exist and can be seen in the western parts of Lekki. Eko and Kuramo were separated by Odo Alarun (known as Curtis's Passage in 1852 and later as Five Cowrie Creek).[25] Kunle Akinsemoyin and Alan Vaughan-Richards attribute this new name to the cost of the ferry from Eko to Kuramo, and nineteenth-century maps show the ferry's location at the northwestern tip of Kuramo.[26] Eventually, British administrators built a bridge across the creek in the 1860s. Kuramo Island itself was split into unequal parts by Igboṣere Creek, which led into the Kuramo Lagoon (known now as Kuramo Waters). Small settlements west of this creek included Magbon, Iperu, Iru, and Ikoyi; east were Igboṣere village (on the banks of the creek) and Maroko village. Several other villages are marked on Speeding's 1898 map, *Lagos Harbour,* but were not named.[27] Another fix-and-fill map from 1932 shows the extent of swamp areas still unclaimed on Victoria Island.[28]

The points of intersection of these streets were also of religious importance; the ara Eko, or inhabitants of Lagos, were often seen leaving offerings and gifts to their orișa at the intersections. A common phrase, "Itta mẹtta ko kọnnọ ẹbọ" (The junction of the roads does not dread sacrifices), shows the importance of this practice.[29] In fact, this practice was so common that Rev. Samuel Ajayi Crowther used it in an example to illustrate the definition of *ita,* or street, in his 1852 dictionary, the *Vocabulary of the Yoruba Language.* When the ara Eko used the word *ita,* they could be referring to a street, an "open place in front of a building," or even just the "open air." So, for instance, Ita Balogun could refer to Balogun Street, Balogun Square, or both.

A WALKING CARTOGRAPHY: S'A TI PENI SỌ ADUGBO ATI ITA L'ORUKỌ

These street names are memories and history now, but at the time they referenced the present. The historical street names and the process of mapping them also provide a spatial vocabulary for writing about historical Lagos through a contemporary lens. For instance, when I write that Faji Market was at the intersection of Isalẹgangan and Massey or that the Iga Idunganran sat between King Street and Great Bridge Street, map 1.2 helps to locate these places in the city. Many former colonial cities in West Africa have naming schemes that inscribe the ambition and scope of colonial power. Street names in Lagos provide a unique opportunity to access an indigenous past usually characterized by archival silences, exploitation, or violence (following the work of Michel-Rolph Trouillot and Marisa Fuentes). Streets in Lagos Island are sites of historical concern, as some of the names used invoke even seventeenth-century narratives. This shows the depth of historical knowledge embedded in the urban fabric. This book presents the first attempt to consider the streets in the early colonial and then independent city as a whole, and to transcribe and interpret the history of the city that is embedded in the expansion and naming of its road network. In this way, I respond to the call of scholars to decolonize archives and knowledge production, with an eye toward privileging perspectives that have been silenced and obscured. As the names of streets in Lagos change over more than a century and a half, they announce new complexities and patterns in this public archive of place. Contrary to historical interpretations that render the physical city as a blank space, featureless until annexed by the British, this chapter pushes audiences to imagine and engage with the old African city, and offers a variety of visual and textual narratives as a basis to do so.

Now, in the twenty-first century, there are hundreds of names in use in Lagos, and they stretch across the entirety of Lagos Island, from Isalẹ Eko in the west to Banana Island in the east. If you factor in Greater Lagos, which includes

FIGURE 1.3. Streets in old Lagos. The historical map is a detail from *Plan of the Town of Lagos—West Africa (in 15 Sheets)*, 1891, 1 inch to 88 feet (scale), 63 × 90 cm, CO 700/Lagos 14, courtesy of the National Archives, Kew, UK. Photographs by Lekan Adedeji and author. Graphic by author.

the mainland, Victoria Island, Lekki Peninsula, and so forth, the names likely number into the thousands. But by mapping Lagos and producing street transcripts as a source, one can decipher the histories of, for instance, silent and sometimes unnamed women who have no place in the archives of early colonial Lagos. The histories of street names in 1860s Lagos are a way to recover silences in the archival and visual record. These street names can function as a literal transcript of stories that escape the traditional archive; analyzing them yields even more narratives of the lives of the ara Eko in the narrow streets of the northwestern part of the island. The ara Eko are invisible in the archive, but by analyzing the histories of streets, we find them and their city. (See figure 1.3.)

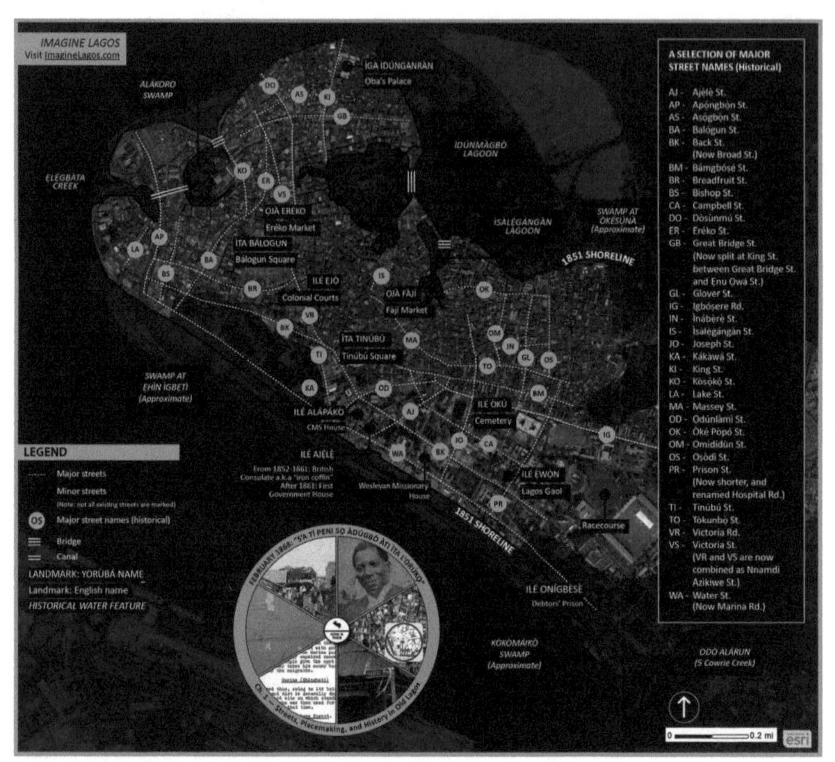

MAP 1.2. "February 1868: 'S'a ti Peni sọ adugbo ati ita l'orukọ.'" The title means "Payne named the neighborhoods and streets in Lagos." Base map courtesy of ESRI. The spelled-out forms of the abbreviated street names are found at the very end of the "A Walking Cartography" section in this chapter. Images in the cartouche are, clockwise from top right, as follows: (R1) portrait of Adeoye Deniga, in Isaac B. Thomas, *Life History of Herbert Macaulay, C.E., Third* (Lagos: Tika-Tore, 1946); (R2) Spyglass over the junction of Omididun Street and Inabẹrẹ Street in Lagos Island; (R3) Inabẹrẹ Street, Lagos Island, 2018, photograph by Lekan Adedeji; (L3) a detail from Adeoye Deniga, *Notes on Lagos Streets*, 3rd ed. (Lagos: Jehovah Shalom / Tika-Tore Printing Press, 1921); (L2) a historical map, J. Hunter, *West Africa: Lagos (Now in Nigeria); Plan of the Marina Area, Showing Streets, Blocks of Buildings and the Owners of Property,* April 13, 1871, 1 inch to 66 feet (scale), MR 1/389/4, courtesy of the National Archives, Kew, UK; (L1) a postcard of Rue Massey, Lagos Island, property of author. Map by author.

Historian Jessica Marie Johnson deploys the concept of null value to demonstrate how Black subjects like the ara Eko are often nullified from archival records, not because they are insignificant, but because they are specifically and deliberately written out of the official record, except when they test the limits of colonial power.[30] Finding and then mapping the nineteenth-century street network does two things to address this nullification: first, by drawing old streets over new ones on a contemporary Google satellite map, it makes the past legible in a contemporary city; and second, it provides a place—specifically in this case, intersections—where Lagosian histories of those who practiced orișa worship, or gained their freedom from domestic or transatlantic enslavement, can be anchored.

The internal logic that organized people, places, and privilege was not apparent to observers until the streets all over the island were paved and named in the wake of the new colonial administration. Lagos's reputation within the British empire was based on the perceptions of these outside spaces. Lynda Nead, an art historian, has shown that streets are "the most visible sign of [a city's] progress or degeneration."[31] Much of the discussion around what Lagos Island lacked in order to thrive as a modern metropolis was framed around the discourse of sanitation, and a significant portion of this narrative played out on the streets of the city. The ambivalent responses of visitors and residents alike to the state of the island and city are well-documented.[32] One missionary, for instance, wrote in early 1852 that Lagos was then an "excellent and well-built native town."[33] But for a long time, Lagos's streets were framed as a problem. Too narrow, too unstable, and too crooked—they were the target of frequent complaints from new residents in the city.

To map the street names and lines from the mid-nineteenth century, I superimpose the historical names of streets—as well as the historical contours of coastlines and other waterways—on a map of the contemporary city. The city's historical street network emerges from sources created at three critical junctions: in the 1860s, during a period when the names and, therefore, meanings were still being fixed; between 1883 and 1891, when the first complete network emerges clearly on maps; and in contemporary Lagos, when the layered street names signal a longing for the past. Historical maps of Lagos swing between empty and filled, rarely giving a sense of growth or expansion. There are no surviving maps of 1860s Lagos.[34] In the 1860s, the city was still confined to two islands: Eko and Ido. Most of the settlement on Eko was in its western half, as the eastern portion was largely a mangrove-lined swamp, similar to Kuramo Island to its south.

From the data collected from 1860s, 1880s, and contemporary cartographic sources, I draw three new maps of Lagos. The first one, "S'a ti Peni sọ Adugbo ati ita l'Orukọ" (see map 1.2), is a map of all the street names that have been documented as being in use before 1872. I have counted about 142, of which nearly three-quarters are in Yoruba. There were two King Streets, and two Victoria Roads in Lagos Island. In the northwest, in the nineteenth century, the longer and more obvious King Street spans the path between the King's Arch at Great Bridge Street and continues north past the palace to the edge of the shore. The shorter King Street (for Dr. Nathaniel King) is in the Faji quarter, east of Tinubu Square, nestled between Ọbadinọ Street and Tinubu Street.

I have found no existing record of the process of the naming of the streets. Even though Otonba John Payne took credit for the names of the streets and listed them in his almanacs, one letter writer in the 1860s gave Administrator Glover credit for the naming scheme.[35] Some names, like Water Street, were in use since the 1850s, while quarter names like Portuguese or Brazilian Town were in use before then.

The street names in use in mid-nineteenth-century Lagos were most likely the result of a dynamic naming process. Three people provide the details of what we know about the development of streets and quarters in 1860s Lagos: Herbert Macaulay, "Professor" Deniga, and Payne. In the nineteenth century, Payne, the city's registrar, was the foremost authority in documenting Lagos. Between 1874 and 1894, he published seventeen versions of his *Lagos and West Africa Almanack and Diary,* an encyclopedic guide containing, like an almanac, a detailed diary, census information, names of jurors, locations of markets, et cetera. Most importantly, he produced a historical timeline of the city, and a list of almost all the city's street names.[36] Historian Liora Bigon identified three categories of street names in Lagos: Yoruba names of people, sites, and names devoted to a handful of colonial officers.[37] In addition to these types are names that reflect a nuanced pattern telling stories about race, diaspora, gender, place, and other themes that are not remembered outside of the street name.

The title of map 1.2 translates to "Payne Names Lagos's Streets and Districts."[38] Payne named sixty-nine streets, six residential districts, and eight landmarks in his almanacs.[39] I have selected and mapped thirty-two major street names that were in use in mid-nineteenth-century Lagos, as seen in map 1.2. In this context, these names are "major" in the sense that they help in understanding the new geographies of the city. In alphabetical order, they are as follows: AJ, Ajẹlẹ Street; AP, Apọngbọn; AS, Aṣọgbọn Street; BA, Balogun Street; BK, Back Street (later Broad Street); BM, Bamgboṣe Street; BR, Breadfruit Street; BS, Bishop Street; CA, Campbell Street; DO, Dosunmu Street; ER, Ereko Street; GB, Great Bridge Street (later split at King Street where it becomes Ẹnu Ọwa Street); GL, Glover Street; IG, Igboṣere Road; IN, Inabẹrẹ Street; IS, Isalẹgangan Street; JO, Joseph Street; KA, Kakawa Street; KI, King Street; KO, Kosọkọ Street; LA, Lake Street; MA, Massey Street; OD, Ọdunlami Street; OK, Oke Popo Street; OM, Omididun Street;

OS, Ọṣodi Street; PR, Prison Street (now shorter and renamed Hospital Road); TI, Tinubu Street; TO, Tokunbọ Street; VR, Victoria Road; VS, Victoria Street (VS and VR have been combined into Nnamdi Azikiwe Street); and WA, Water Street (later, Marina Road). These major street names cover the six original districts in Lagos—Iga, Ọfin, Olowogbowo, Itolo, and Oko Faji—as well as the new ones, such as Ẹpẹtẹdo, Oke Popo, and Oke Itẹ in the east. The "minor" street names are also historically and culturally significant, (and are more numerous,) but are often harder to document. They feature names like Kosseh, Tijuiku, and Alof.

ROAD BUILDING IN OLD LAGOS: WATER STREET AND BACK STREET

Streets were also sites of contestation. Tensions over old and new, between the emerging "modernism" of colonialism and "backward-looking" sense of the ara Eko, were negotiated in the streets. The evidence around the road construction in the 1860s illustrates how the indigenous inhabitants were attached to their ways of life, especially in the way they protested British ambitions for road building. Analysis of the encounters around the construction of Water Street and Back Street offers much information about the histories of midcentury Lagos. Water Street, directly adjacent to the river, was forty feet wide. Broad street, at sixty feet wide, was imagined as a convenient buffer between the European merchants and the African population in Oko Faji and Ereko to the north and Olowogbowo to the west.

In the 1850s, Lagos was a rapidly growing society, with an influx of formerly enslaved peoples from Sierra Leone, Cuba, and Brazil and from towns surrounding Lagos. To the immigrant Saro and British, it was clear what needed to be done on the island, and around it, to overcome the frequent fires, the geographical setbacks of the Lagos Lagoon, and the sandbar that separated the island from the sea. They wanted a "modern" city with the creeks filled in, roads widened, straightened, paved, and named, and housing material changed from thatch and mud to tin and brick. However, the ara Eko did not always agree with the proposed changes, and their resistance to these interventions widened the distance between the indigenous quarters—Ọfin, Ereko, Olowogbowo, Itọlọ, Faji, and Iga, which were now styled as "Old Town"—and the rest of the island, provoking a divide between the ara Eko and those in the so-called "civilized" parts of the city.

The British annexed Lagos in 1861, after a ten-year consular period following the successful bombardment of the city. In 1863, Governor Henry Stanhope Freeman estimated that the population of Lagos had grown to forty

FIGURE 1.4. Population of Lagos in the nineteenth century. Photograph is of Balogun Square, Lagos, undated, CO 1069/71/86. Courtesy of the National Archives, Kew, UK.

thousand, with an additional twelve thousand people rounding out the colony in Badagry, Palma, and the villages between them.[40] (See figure 1.4.)

The 1860s was a time of new infrastructure all around old Lagos. Lighting the streets alone cost £200 by 1871. A wide canal was cut across Ọfin to link Alakoro to Ẹlẹgbata Creek, and bridges were built across Idunmagbo and Isalẹgangan Lagoons, and Five Cowrie Creek. Great Bridge Street connected Isalẹ Eko to Ido. In addition to streets in Eko, new streets were laid out on the mainland: Robert Campbell, as acting colonial surveyor, laid out a new town there, complete with street names as well.[41]

To walk in colonial Lagos was a very deliberate gesture: walking was avoided by those who could afford not to. In a city with few tarred or even paved roads, where many avenues seemed (to visitors) unintentional and not intended for foot traffic, the preferred modes of transport were canoes via the lagoons and creeks. Wealthy Lagosians were often spotted in their large flotillas circling the island, or riding in town on horseback, while the poorer inhabitants had to make do with navigating the labyrinthine roads, avenues, and frequently intersecting passageways. The town's dearth of concrete infrastructure—compared to other colonial cities like Bathurst (with its wide roads and, in some areas, grid-like streetscape) and especially to

London—was keenly felt. Some accounts of the streets were less kind. One visitor in 1859 wrote that the city, like other "native African towns has no streets such as would be called in a civilized country."[42] Walking around the town was avoided if possible (according to some Europeans); thus, the construction of wider paved thoroughfares beginning in the 1860s had an immediate and dramatic impact on the southern edge of the city, where most of the European and Christian enterprise was located. Nowhere were these construction-related disruptions clearer than in the disputes over burial practices and the destruction of houses to widen roads. In 1862, the colonial government had estimated that it would spend around £560 on the construction of roads, streets, and bridges. Instead, it spent much more, with just over £869 going toward covering the costs of "houses pulled down" and compensation for the owners.[43]

Richard Burton, the infamous explorer, observed the city while aboard William McCoskry's ship as they sailed into Lagos in July 1861 and his often-quoted description of Lagos's southern edge talks about the new buildings on the island from the Five Cowrie Creek, which framed the island's east edge, to Olowogbowo, the furthest district to the south (see figure 1.2). He described the "thin line of European buildings" as the only ones having any merit and pointed out that they were backed by "a large native town, imperceptible from the sea, and mainly fronting the Ikoradu [*sic*] Lake."[44] The newly constructed road along this edge, known at the time as Water Street, was deemed the only pleasant walk in the city.[45] He cataloged a long list of necessary improvements. Perhaps it was because of Burton's advice that, five months later, by October, Acting Consul McCoskry finished the construction of the promenade on the south side of the island, referred to at the time as the Esplanade.[46]

The Ẹhin Igbẹti area was swampy and covered with a thick brush before the merchants, missionaries, and consuls moved there. Lagosians saw it as the back of their city; it was miles away from Isalẹ Eko and the inhabited districts in Lagos. The new European settlers saw advantages in this distance and made plans to redevelop the area to their liking. Reflecting on the new district in the 1870s, Administrator Glover described it as the "receptacle of public filth, and the hiding place of thieves when pursued."[47] This was the reason they remade the district: it was done in the interest of "public health and security."[48] However, there were other reasons: this southern edge of the city faced the Atlantic, and so was the perfect part of Lagos to establish commercial activities—warehouses and factories to store and ship goods from the interior—and also to be as far as possible from the indigenous Lagosians. The residential quarters, from Isalẹ Eko to Faji, were densely populated. In addition to building Water Street along the Lagos Lagoon, another broad boulevard—Back Street—was built behind it as a physical buffer between Europeans and the Africans.

Despite ambitions of developing a grid pattern like that in Freetown, it only happened in the Ẹhin Igbẹti area, which was mostly uninhabited in the 1850s. Between Marina Street and Broad Street and west from Apọngbọn Street to Prison Street is the only place where the grid exists. The rest of the streets in Eko conform to the different, more organic pattern that likely reflects the physical, social, and cultural geography of the island and city.

British administrators in Lagos made two specific interventions in road building in the early 1860s. The first was the construction of Water Street along the Atlantic side of the island in 1861, undertaken by McCoskry. The improvements in the technology of street making did not go unnoticed. "We are glad," noted an editorial in *The Anglo-African,* "to see in every direction evidence of the determination to introduce civilized improvements into the town. The new street in the east end, parallel to Water Street is steadily progressing."[49]

Clearly, Lagos was still framed as an urgent problem that had to be solved. By 1863, Water Street, Balogun Street (also known as Ita Balogun), Breadfruit Street, Balbina Street, Bankọle Street, and Broad Street had been named and paved.[50] By July 1866, several existing streets had been widened, paved, or newly constructed.

Glover began several infrastructure projects in 1863. Nearly £2,000 was spent on repairing the Government House, sinking new wells, making progress on the "partially erected" Court House in Tinubu Square, and building the Customs House.[51] The construction of new roads and expansion of existing ones nearly exceeded the original estimate by 50 percent. Most of the cost likely went to compensating people whose land was taken and whose homes had to be destroyed to make way for new and improved streets. A new bridge had to be built over Alakoro Swamp to connect it to Itọlọ. This meant that it was now safer to cross between the two areas and was most likely the first of three Bridge Streets in old Lagos. In the east, new bridges were also built over Idunmagbo Lagoon and Isalẹgangan Lagoon, connecting the Isalẹgangan district to Idunṣagbe in the west and Oke Popo in the east. By the early 1870s, the cost of compensating people for their lost homes and land was almost £850, and another £179 was spent on constructing a new road in the eastern district.

It seemed that the budget for building roads increased every year. In 1867, the government spending on roads peaked at £2,192, a dramatic increase from the £147 spent in 1866.[52] Of that money, £1,023 was required for a military expedition, which needed a road "through the western district,"[53] while the rest was used to continue widening the streets and repair fire damage.[54] In 1868, expenditure fell by around £95 because less compensation was needed to pay for the houses that were torn down.[55] By 1871, cleaning the streets cost an estimated £285 a year, while road making cost around £811.[56] Repairs to bridges were estimated at around £50, with Idunmagbo Bridge talking up 20 percent of the funds.

Despite how much the new construction was celebrated, there were still complaints. "Vox Populi" wrote on December 4, 1871, that the rainy season exposed the weakness of the street infrastructure: "The streets of Lagos during the rainy season testified to every spectator, white or black, that whatever else he may have known, Mr. John Hunter [the colonial engineer] does not know what is called drainage."[57]

By 1868, the Iga quarter in the north had been transformed into four different neighborhoods: Idunṣagbe, Idumọta, Ebute Ero, and Idumagbo, their names reflecting existing sites of historical importance. In 1869, Glover reported that "wide streets are cut in all directions, and principally in the one calculated to allow the wind from the lagoon to cross the island."[58] Scavengers were employed to clean the streets and wharves daily.[59] The new roads centralized the sale of firewood, moving this business away from the city's edge, and into newly marked spaces. Glover believed the effect of these sanitary activities reached so far it could even shift people away from their traditional beliefs. The staff surgeon reported that home burials were "entirely done away with" in 1869.[60] In 1872, the government spent £850 to compensate people whose homes were torn down to repair and improve existing roads, and spent £179 to create a new road to the "Eastern District," likely the extension of Igboṣere Road.[61]

Despite these minor improvements, the idea of Lagos being undesirable persisted through the 1870s and well into the 1880s. "Some parts of the town are so offensive," wrote one resident, "that it is hardly possible to get through it without a painful effort."[62]

CONTESTING PLACE AND POWER: THE OLD CEMETERY AND BANNER BROTHERS

After the cession in 1861, the new British administration in Lagos claimed ownership of all the land in Lagos, from the interior to the foreshore. This was challenged immediately, both by the Lagos chiefs (who had ancestral claims to the land) and the new European merchants (who had title to property). The new administration's treaty with Ọba Dosunmu indicated he had given up all claims to the island. The "white cap chiefs" (as they were colloquially known due to the headgear they favored) claimed Dosunmu could not have properly given away the land, as they had claims stretching back decades, if not centuries, which established them as the actual owners of land in Lagos.

The development of the new Christian cemetery reveals how Eko was viewed as a physical, social, and spiritual space. Benjamin Campbell, the British consul, commented on the increasing alienation of the ara Eko. In the colonial report for 1863, the governor noted the ill effects that the change to British rule had wrought, writing that "British law, no matter how mildly and considerately administered, had struck a deathblow at the many of the habits, customs and prejudices of the inhabitants of Lagos."[63] However, he neglected to add the disruptions the merchant community on the riverside experienced.

The trouble began when Glover, the lieutenant governor, announced his plan to widen the streets in the Iga quarter and build a new market shed. This plan was bitterly opposed, and the locals reacted with the "deepest horror" when the colonial surveyor and his men arrived to measure and plan these interventions.[64] Though the objection of the chiefs and people were at first dismissed, on later reflection it was found that the objection was because the plan would lead to the destruction of houses where people had buried their dead for generations, severing ancestral links.

Nevertheless, the Lt. Gov. Captain John Glover attempted to enact an ordinance to ban any future burials within compounds, insisting that the Lagosians use the burial grounds in "the fields" east of the city. Forcing people to bury their dead outside the home only exacerbated the issue. The ọba, in a letter responding to the new government's request, wrote that from "time immemorial" and "in common with all the natives throughout this section of Africa," they had always buried their dead in the homes where they had lived. They desired to maintain continuity with their history, and further, it was considered "the greatest dishonour that could be offered to the memory of the dead to bury them in the fields."[65] And in refusing the new market, the people *insisted* their streets were adequate, and that the locations of the markets in Ebute Ero, Iga, and Ereko were sufficient. Such analysis of local responses to "improvement" schemes reveals the symbolic value people assigned to spaces.

While street building in old Lagos is often seen as generating conflict between the administration and the Black communities, there is evidence in at least one case where a new road caused tension between the administration and the merchant community. The correspondence from the controversies over one plot of land on Water Street illustrates some of the conflicts that arose with European merchants as well. The Banner Brothers' 1860s land complaint reveals more about the colonial government's plans and the significance of the redevelopment of Ẹhin Igbẹti.

The Banner Brothers—Edward, Gregson, and John—were among the earliest arrivals in Lagos after the bombardment in 1851. Less than a dozen European merchants set up at Lagos, with some coming from nearby Badagry. The brothers owned and operated one of the oldest European concerns in Lagos with connections to legitimate commerce. They bartered British manufactured goods for commodities from the interior.[66] They had factories and dwellings at both Badagry and Lagos, and in each place they intentionally chose and requested plots of land by the water to store and transport their cargo. From Lagos, they used the Ogun River to move commodities to and from the interior.[67]

In Lagos, they built their factory at the intersection of what became Water Street and Bishop Street (see map 1.2). They had received a land grant from Ọba Dosunmu dated February 15, 1856, in the Ẹhin Igbẹti area, through their agent Amice Le Gresley. In it, Ọba Dosunmu agreed to give them land at Ẹhin Igbẹti (written as "Ebete") between the properties of Henry Pratt and S. B. Williams.[68] They were given 112 feet of water frontage, with 120 feet in the rear.[69] Dosunmu's seal was affixed to certify the transaction.[70]

The new British government claimed the foreshore in order to construct Water Street, which had the effect of cutting off the Banner Brothers from their jetty and wharf. They clearly needed a jetty for "landing, storing and shipping," and had invested thousands of pounds to secure and build their factories. Without this water frontage, they wrote, their investments in both towns would be a "complete waste." The Banners said the government was "pretending" that the land belonged to it. Much like the indigenous Lagosians, they also planned to "resist to the uttermost" the government's claims.[71] The administrators insisted their grants measured their plots inaccurately and, therefore, the government could take the land.

The Banners claimed to have planted trees along Water Street on the lagoon side to mark a boundary, restricting access to their boat house and fire engine, which was the only one in the city. Glover addressed their complaints in an August 1871 letter.[72] He objected that their claim to the land lasted forever. The Banners had received a Crown grant in 1865, but Glover argued that the new dimension in the deed did not match their plot. Both sides had the plot remeasured. Only the intimate map from the government archive is available.[73] Glover said the government had claimed the space where the Marina was constructed, and that the Banners had full use of the waterside for their boathouses and jetty, but they could not exclude the public from using the road. "It must be considered," wrote Glover in his response to their complaints, "that this 60ft street was made in the interest of Europeans, and for the protection of their properties, and divides or cuts off the European Portion of the town from the Natives. None should be more alive to the advantage of this, than the Messrs. Banner, who were greater sufferers from fire than any other European in Lagos, prior to the sixty feet street having been cut."[74] Eventually, their complaint was resolved, and the Banner Brothers accepted compensation.

FROM BREADFRUIT TO AṢỌGBỌN

If Breadfruit Street could speak, it might talk of the histories of slavery it witnessed in the city. Before the trees that it was named after were planted, missionaries described it as a site of slave barracoons, and of a whipping post for enslaved people. After the bombardment, and with the coming of Christianity, it witnessed a dramatic overhaul. As Charles Gollmer noted in his journal for 1852: "We have pulled down their Barracoons, let us give them churches instead."[75]

Just before midday on Monday, June 14, 1852, two men walked together in the streets of Isalẹ Eko, the historical core of mid-nineteenth-century Lagos. Gilbert Lawson had been enslaved from Lagos and rescued on the open seas by the British Navy's Preventative Squadron. James White had come to Lagos from Freetown via Badagry.[76] Both walked together in the city as free men. It had been six months since the British Navy's bombardments of Lagos had destroyed half the city and exchanged one king for another. Ọba Akitoye, the former king, had promised to abolish slavery and keep Lagos for the free. In December 1851, he had joined forces with the British to seize his throne back from his nephew Ọba Kosọkọ.

Lagosians were slowly rebuilding their city from the rubble. That Monday, the men were on their way to Chief Aṣọgbọn's *Iga* (palace), which was well situated along what would become Aṣọgbọn Street in the Idunmọta district. White was a Black catechist in the Church Missionary Society, and he walked in the city often. He wrote that he enjoyed getting to know Lagosians and their city, while looking for opportunities to proselytize.

That morning, as he accompanied Lawson through the streets of Lagos, they spoke about their shared past. White was likely only one generation removed from slavery and, like Lawson, had come to Lagos after the bombardments. But Lawson had a different goal that morning: he longed to confront the chief who had enslaved him and sold him away from Lagos.[77]

White had encountered men like Lawson in Lagos who even after several years could "point [a] finger to the very house from which he was sold."[78] Clearly, some parts of Lagos were etched indelibly in their minds. Every street, every barracoon, and every Iga held painful memories for some, while also representing political and cultural power for others. Along the way, Lawson told White the story of his capture and sale. It had happened early one morning "when he least expected it,"[79] compounding his tragedy. But he had been rescued along the way by the Preventative Squadron and taken to Sierra Leone. How he longed to confront the man who had tried to seal his fate! Soon they arrived at Aṣọgbọn Street and sat patiently in Aṣọgbọn's house, waiting for his servants to fetch him.

Aṣọgbọn, an important *abagbọn* (war chief), was well acquainted with White but did not seem to recognize the man he had sold. "I am the man who some years ago was your slave, and whom you sold to one of the Portuguese," Lawson said. He added the details of his rescue and how his life had changed in Freetown. "Little did you think of seeing me again," he continued, "when that morning you ordered me to be sold to [them]." The chief was certainly not expecting this encounter. On being confronted, Aṣọgbọn became very agitated. He tried to redirect the conversation, but Lawson would not be discouraged. He persisted in his questioning. White, describing the scene later in his journal, wrote, "The confusion into which the chief was thrown, can be better imagined than described."[80]

In his 1961 dissertation, Akin Mabogunjẹ created a map of the residences of twenty major chiefs in old Lagos, or Eko. According to his research, the home of Aṣọgbọn (one of the six abagbọn chiefs he mapped) was in the Idunmọta quarter. A closer inspection of his map puts this location on Aṣọgbọn Street, likely at its intersection with Dosunmu Street, or with Victoria Street (which puts it at the edge of Idunmọta).

WRITING GENDER, MONEY, AND FREEDOM INTO PLACE

"The real Lagosian," wrote Rev. J. B. Wood in his 1870s account of the history of Lagos, "loves above everything else to be a trader."[81] Thus, it is only fitting that so many of the men with streets that bear their names were traders, with Ṣitta, Sogoro (Sogunro), Abisogun identified as "principal native traders in Lagos."[82] A handful of European traders are remembered as well, such as Regis Aine, Balbina, and Giambattista Scala (who even had a square in Faji named after him).[83]

There are only three examples of women written into space in old Lagos. Outside of recognizing Tinubu, Tijuiku, and Faji, the names of the streets make the city an entirely masculine space. At least two important women—Tinubu and Tijuiku —were exiled from Lagos in the 1850s and 1860s. One man, Bamgboṣe, was also considered for exile, but the colonial administrators forgave him and allowed him to stay. Deniga described him as an "idolator from Brazil who worshipped Sango."[84] Each of their stories is written in the physical history of the city through street names in Oko Faji, Brazilian Town, and Agarawu.

Some accounts say Madam Tinubu arrived in Lagos in 1835 as one of Prince Adele's wives, while some say she arrived later. She left for Badagry a few years later, but returned after Ọba Kosọkọ was dethroned in 1851. She was known for her business acumen, and often functioned as a sort of intermediary between the Lagos traders and the European merchants. British colonial officers referred to her as "the woman Tinaboo." Several men in town seem to have conspired against her. Eventually, Tinubu was banished from Lagos on April 15, 1856. According to one historian, "The cause [of the jealousy] was this woman was very rich and her wealth was so great that she hardly knew the number of her slaves"; this is a dubious claim.[85] She is said to have owned most of the land in present-day Tinubu Square, which was named after her, and where a statue dedicated to her still stands.

In an 1853 note Benjamin Campbell wrote, "There is another mischievous person in Lagos whose removal is very desirable, but I fear difficult to effect, the woman Tinaboo [Tinubu], the late Akintoye's niece. She is heavily indebted to some merchants here and she will not pay them."[86] "This woman is the terror of the place," wrote Consul Louis Fraser in 1853. She is the one that "kings and chiefs are afraid of!"[87] Yet, her name adorns one of the most important squares in Lagos Island, and is the name of a small street as well.

Ita Onibaba was said to be a section of Victoria Street and was named for Tijuiku ("shame death" in English), a woman described as a high priestess in the worship of Ṣọpọnọ, the Yoruba god of smallpox. Ṣọpọnọ Street is a short street in the Agarawu area, and it intersects Victoria Street.[88] Tijuiku was expelled from Lagos in 1869, at the request of the colonial administration after allegations that she was enriching herself at the expense of vulnerable Lagosians who died in alarming numbers from small pox. Her belongings were put in Tinubu Square and burned.[89]

These streets clearly articulate the histories of people of African descent from around the diaspora who made an indelible impact on the space. For instance, Joseph Street reminds us that for a few brief years in the 1860s, Joseph Harden, an African American missionary, made Lagos his home. Until his untimely death in 1864, he was also the first brickmaker in Lagos, a feat that meant that constructing homes of more durable material than mud became possible for more people. Even his "broken bricks" were for sale, emphasizing the growing need for these kinds of materials in the city, even at a reduced quality.[90]

I have identified ten street names from the 1860s that were named after men who were previously enslaved. The origins of some names are well known, like Bishop Street, but for the more obscure ones, I have relied on interpretations by Deniga, whose *Notes on Lagos Streets* (first published in the late nineteenth century) suggests the meanings of some of the street names.

Some street names, like Tokunbọ (which means "born overseas") and Bamgboṣe in the Brazilian quarter, were unsurprisingly named after men enslaved in Brazil who later returned to Lagos. In the west, mostly in Olowogbowo, are streets named after Saro men: Bishop Street and Martins Street (for Rev. Samuel Crowther and possibly Rev. Martins, respectively). In Faji and Agarawu are names like Labinjọ, Akani, and Ọbadinọ, names that Deniga attributes to men who were enslaved locally but who eventually earned their freedom. Unsurprisingly, all of these names are in use outside of the ọba's quarter.

~

> The town, however, and the townspeople as well, wore a new and greatly improved appearance, the work of the great benefactor of West African cities, "General Conflagration." Three fires had followed one another in regular succession through November, December, and January, 1863; and the fire god will continue to "rule the roast" till men adopt some more sensible style of roofing than thatch and "Calabar mats."
>
> —Richard F. Burton, *A Visit to Gelele, King of Dahome*

While it is clear that streets and histories, and the past and present, intersect in old Lagos in interesting ways, walking through the streets and returning to the archives led to more questions than answers. Streets are the keepers of stories of the silenced ara Eko and ọmọ Eko, and at no other time was this more obvious than in the 1860s, when the streets were widened and named.

Perhaps the label "King Street" was an appeal to the timelessness of the institution, but Akitoye's omission is a glaring oversight in a naming scheme that focused on writing masculinity and power into place. Reading the streets this way shows how fires, when archived in space, have the potential to highlight the past. Fire was the most expressive version of spontaneous and vindictive actions in Lagos, and it was also no respecter of the social or cultural divisions that were the markers of the newly developing residential quarters in Lagos.[91] Fire, whether the result of careless cooking, accidents, or fireworks, or even deliberately set to signal the beginning of a political attack, engulfed entire houses, and sometimes neighborhoods, and crossed lines between the Saro, ara Eko, and other populations, checked only by natural gaps that were a result of the many inlets and creeks that interrupted the Lagos shoreline.[92] The *ina ayọrunbọ* of 1859 was a conflagration that made the ara Eko feel the sky was crashing down upon them, and the fire at the consulate's barracks (sparked by an accident with ammunition) was equally destructive, and spread outside of the compound, leaving several houses destroyed in its wake.[93] In 1853, the fires set during Kosọkọ's attack burned houses from Ebute Ero to Faji, and a few years later, in October 1859, around fifty houses in Iga quarter were burned. Stopping these frequent fires was one of the first priorities of the island's new owners.

Again, in Lagos, we find that the documentation of destruction allows us to reconstruct city life. It is through the fires that destroyed houses and spread between quarters that we learn of the proximity in which people lived, how they slept, how they cooked, and how they lived their lives. We find that in Idumọta, women made a living by frying fish in the streets, and that too could cause a large conflagration.

Very few parts of the city were safe from fire, since so many of the inhabitants still roofed their houses with thatch, despite the proclamation in 1863 that everyone had to use the less flammable Calabar mats. Ordinance number 8, issued in April 1863, was drafted for "the better protection of the Town of Lagos from fire."[94] In 1894, Payne recorded the most devastating fires that had plagued the city between 1859 and 1892; he counted at least thirty-nine fires that had catastrophic results. There were at least five fires where at least three hundred houses burned; in ina Ọbun Eko, in 1877, and ina Ita Bamgboṣe, in 1873, at least five hundred houses were damaged. Some of these fires spread along streets—like, in 1873, when one fire spread from Ita Tinubu to Victoria Road—as well as between quarters. In 1884, the ina ita Olowogbowo spread from Chapel Street all the way north to Isalẹ Eko.

The fires also destroyed important landmarks: the Ebute Ero Church in 1859 and a mosque in Ita Shitta in 1875.

Many lives were lost in the eastern quarters in Lagos, as the ina Ita Bamgboṣe spread to Oke Popo in 1876. Five years later, a child was killed in the ina Ofin. In 1891, a blind woman was burned to death as the fire spread from Ẹnu Ọwa to Idunmagbo. Of these "great fires," at least two happened in the period under consideration for this chapter: Payne records the ina ayọrunbọ on March 4, 1859,[95] and ina Ebute Ero that destroyed the church on October 22, 1868.[96]

Children and the elderly often suffered the most from the fires, especially from the effects of smoke inhalation.[97] An *African Times* editorial and letter provided vivid accounts of the large fires in 1862 and 1863.[98] A whole family was found dead in July. It had been a rainy night, and they fell asleep in a damp, windowless room with a fire burning. A similar occurrence had unfolded in Abẹokuta. That same year, another fire raged through the Faji Market area.[99] The fire broke out around eight on the night of Saturday, October 21, 1865. There were reports of a strong wind that spread the flames from roof to roof; the flammable thatch acted as an accelerant for the fire. Luckily for the residents, men from the Hausa Police Station were soon on the scene and helped to check the flames quickly.[100]

What does it mean that the itinerary of a fire is contained within the street names of Lagos? Our journey in the Ẹpẹtẹdo neighborhood came full circle when we came across Inasa Court. Its name is a tempting bookend to the narratives of fire and fire-mongering in the 1850s and 1860s. The destructive fires in the city had left little physical trace, but have been immortalized in song and in the records of Payne's almanacs. If Inabẹrẹ had suggested the beginning of a fire, perhaps Inasa indicated where the fire had ended. Here, in Ẹpẹtẹdo, was the beginning and end of a fire that has not been remembered consciously. Inasa—which could mean "the fire fled"—possibly marked the end of another catastrophe that may have begun less than a mile away at Inabẹrẹ Street. However, even these transcripts can be misleading and may prompt different readings.

There are no large fires recorded in Ẹpẹtẹdo in the nineteenth century. There is no evidence of Inasa Court in the nineteenth- or twentieth-century maps, but the name occurs in court testimonies in the 1930s. What seemed to be part of the narratives of fires in Lagos turned out to more likely be a reference to a compound owned by the Inasa family, descendants of Inasa (or Nassah), a man who was one of Oṣodi Tapa's many followers and returned to the city with him in 1862.[101] This land was the subject of a series of disputes archived in early twentieth-century court cases over individual versus family ownership of land, and whether Oṣodi's family actually retained the right to profit from the land, even though he had divided the land among his followers.[102]

During a case between Sakariyawo Oṣodi (one of Oṣodi Tapa's sons) and Dakolo, the former testified that Governor Glover had indeed given his father the whole of the Ẹpẹtẹdo quarter, and they understood that the family had retained the rights to sell the land. In resolving the suit, there was some confusion about whether any of Kosọkọ's followers also settled there, or whether the land was solely Oṣodi's to distribute.[103] Regardless, the area was given the name Ẹpẹtẹdo because of Kosọkọ and Tapa's exile in Ẹpẹ after the 1851 bombardment.[104] Oṣodi's son maintained that Oṣodi settled there and built twenty-one compounds, each headed by a different man known as an *oloko*. Inasa Court was said to be one of these headmen, which would explain why the court bore his name.[105] On assessing the effect of Crown grants, the judge in the case wrote that any individual's "executors, administrators, and assigns forever" were bound to introduce the idea of individual ownership and to lead the grantee to lose sight of the original relation between himself and his overlord.[106] In 1869 or so, the Lagos government decided that each compound should have a Crown grant and encouraged each oloko to acquire one in his own name. The case presented a conclusive piece of evidence: a grant for Inasa Court was issued on May 5, 1869, in the name of Inasa.[107]

Even though street names can be misleading, they can still offer opportunities for telling stories that have been ignored. Fires, like bombardments, ironically have a way of preserving significant sites and people. In the streets of Lagos, there are stories that are still missing and still being recovered as the names change. The next chapter looks at how some of these same sites acquired and changed meaning.

2 ⁓ Who Broke Lagos?

At Lake and Ẹlẹgbata

> In the beginning there was a river. The river became a road and the road branched out to the whole world. And because the road was once a river it was always hungry.
>
> —Ben Okri, *The Famished Road*

IN JULY 2018, we (Pẹlu, Lekan, and I) went looking for the outlines of Ẹlẹgbata Creek and Alakoro Swamp on the western edge of Lagos. These two water bodies had been built in the antimalaria and antiswamp campaigns of the late nineteenth century, but held a special historical significance in the preservation of the city of Lagos during the 1851 bombardments. While anchored off the north point of Lagos Island after successfully bombarding Lagos in December 1851, Captain Lewis Jones (of the HMS *Bloodhound*) wrote of how a "creek and swamp" had protected the eastern portion of Lagos Island from being destroyed by the fire during his attack.[1] This striking note is one of the few references to the important water features that marked the edges of Lagos Island, highlighting their significance in saving the Olowogbowo, Itọlọ, and Faji quarters.

Today, the western edge of Lagos Island is nicely rounded; both Adeniji Adele Road and Ring Road sweep north from the roundabout at Apọngbọn Street on the Atlantic-facing shore. They trace the shoreline up and around, passing under Eko Bridge, which connects Apọngbọn Street and the island

MAP 2.1. Spyglass over Lake Street and Ẹlẹgbata Street in Lagos Island. Base map and Spyglass courtesy of ESRI. The historical map is a detail from *Plan of the Town of Lagos—West Africa (in 15 Sheets)*, 1891, 1 inch to 88 feet (scale), 63 × 90 cm, CO 700/Lagos 14, courtesy of the National Archives, Kew, UK. Map by author.

to the mainland, before proceeding northeast past Carter Bridge and finally looping south, back into the heart of the island. But in the 1840s that shoreline was jagged and fragmented, and the landmarks, with one exception, were much different (see map 2.2 but note the inversion). Today, the southward sweep of the northeast end of Adeniji Adele Road marks where the northern shoreline of the island plunged south along Idunmagbo Lagoon, the eastern edge of Isalẹ Eko, which was home to the ọba's palace on the northern shore and was accessible via a bottleneck between the lagoon and Alakoro Swamp to the west.

This swamp also defined one edge of the bottleneck by which Ọfin was accessible; the other edge was defined by Ẹlẹgbata Creek, which used to dominate the west end of the island, dividing Ọfin from Olowogbowo. Map 2.1 is a screenshot of a Spyglass map on Lagos Island in the Ẹlẹgbata area. The Spyglass reveals two mapped layers at once: on top is a satellite map of Lagos, and beneath is a historical map of the same site. In this screenshot, you can see the outlines of Ẹlẹgbata Creek before it was reclaimed in the early twentieth century. Now, in Lagos, Lake and Ẹlẹgbata Streets mark the curves of the former creek.

Ẹlẹgbata Creek was underneath what is now roughly the areas between the western end of Apọngbọn Street and Ẹlẹgbata Street, crisscrossed by smaller streets like Griffin Street and Savage Lane. The intersection of

Ẹlẹgbata and Lake form the easternmost point of the creek, so this is where we started our walk. Ẹlẹgbata Creek is depicted in historical fix-and-fill maps as separating Itọlọ and Olowogbowo.[2]

These streets are reminders of the role the water bodies played in defining the precolonial urbanism of Lagos and protecting the western end of the island from total annihilation by fire during the civil war of 1845 and again during the British bombardment of 1851 when they campaigned to restore an ọba less friendly to slavery. Eko's mid-nineteenth-century spatial politics was shaped, first, with reference not to the Atlantic Ocean, as it would be by the British, but northward to the Bight of Benin and its other urban centers in the civil wars of succession in 1845 and 1851.

This chapter explores the historical genesis of, and cartographic basis for, tensions between Ọba Kosọkọ and Ọba Akitoye's visions for Lagos. As Kosọkọ's city, "Eko" would have retained much of its indigenous gestures to Yoruba urbanism, mixed in with the marks of transatlantic slavery. As Akitoye's city, regained with the help of Christian missionaries and the British Navy, "Lagos" was reimagined as a place where everyone was, or could be, free. The nineteenth-century history of these two-cities-in-one is inscribed in the earliest attempts to record the spatial contours of the city on the eve of its December 1851 bombardment and reduction.

In the first section, I lay out the stakes and sources for spatially reconstructing Eko through the 1840s up to its reduction in 1851. My reconstruction of the pre-1851 city is based on the accounts of the bombardment, using written and mapped descriptions of Eko before and after the shelling, by men who visited the city, surveilled bombing targets, and walked the city with Akitoye to return him to the palace. The city as Kosọkọ ruled it from 1845 to 1851 emerges from the rubble, represented by map 2.2, "Marking Ọba Kosọkọ's Eko," bringing with it insights about the role that slavery, rivalry, and succession played in shaping Lagos. While the chapter begins in the aftermath, in 1852, of the Ija Agidingbi (Booming Battle, or Reduction of Lagos), it then circles back to 1845, retelling this transformation of Lagos resulting from the 1845 Ogun Olomiro (Saltwater War).

But how did regional political ambitions converge at Lagos? In answering this question, the second section, "Eko: Succession, Slavery, and an Emptied City," writes this space into the political history of the region, reconstructing the spaces in which the ara Eko struggled; doing so reveals what was really at stake in the quest for political, social, and economic control of Lagos. It explores the changes in Lagos's urban morphology between 1845 and 1851, linking it to the politics that not only dominated events in

Lagos but also spread to Badagry, Porto Novo, and Abẹokuta. With violent political acts recast as specifically spatial narratives, the section demonstrates how the political and spatial axes are inextricably linked and locally inflected, looking at Eko's succession politics via the imbrication of three spatialities—the Bight of Benin, the island itself, and finally the city's quarters—to reveal what is at stake in reinventing power in place. In the late 1840s, Lagos was mired in a cycle of physical destruction and reconstruction, coupled with the demographic shifts that accompanied such activities. Threats to life and property meant a constant flow of refugees to neighboring towns and villages; in times of relative peace, men and women would return to rebuild their compounds, markets, and quarters.

The question of "Who broke Lagos?" is best answered quarter by quarter in the city, and these responses had lagoon-wide implications, remaking lagoon relationships and tilting power toward eventual British occupation in 1861. These shifts are observed in the symbolic capital of destroyed and rebuilt sites; for instance, how destroying the barracoons, the enclosures confining people being sold into slavery, on the edge of the city destabilized the lagoon commerce in shipping enslaved people; how the development of new quarters and razing old ones reflected entangled fortunes around the Bight; and finally, how British action drove a definite wedge in the political and economic cohesion of this lagoon-bound space.

A WALKING CARTOGRAPHY: MARKING ỌBA KOSỌKỌ'S EKO

"Marking Ọba Kosọkọ's Eko" (map 2.2) is a speculative map based on three different versions of a map of Lagos Island used to plan and then record the attack on Lagos. Each is a perimeter map made during the bombardment of Lagos by naval officers in 1851. Each map varies slightly in its rendition of the town and island, and in its claiming of space and landmarks. In Thomas Earl, J. D. Curtis, and W. Harris's *The Entrance of the River Lagos* (figure 2.1), we see a distorted yet recognizable outline of Lagos Island and the islands and land masses that surround it.[3] This map focuses on the defensive details of the city, outlining where Kosọkọ and his men hid their long guns and ammunition. While the other two perimeter maps, Thomas Earl's 1851 *The Entrance to the River Lagos* and H. P. Ward's *A Plan of the River Lagos,* make visual arguments about Lagos's emptiness as opportunity, Earl, Curtis, and Harris's *Lagos River* shows it as a violent space, with the weapons deployed to protect the business of transatlantic slavery.[4]

This speculative map draws on maps, court records, and reports from the bombardment to reconstruct Kosọkọ's city on the eve of the second

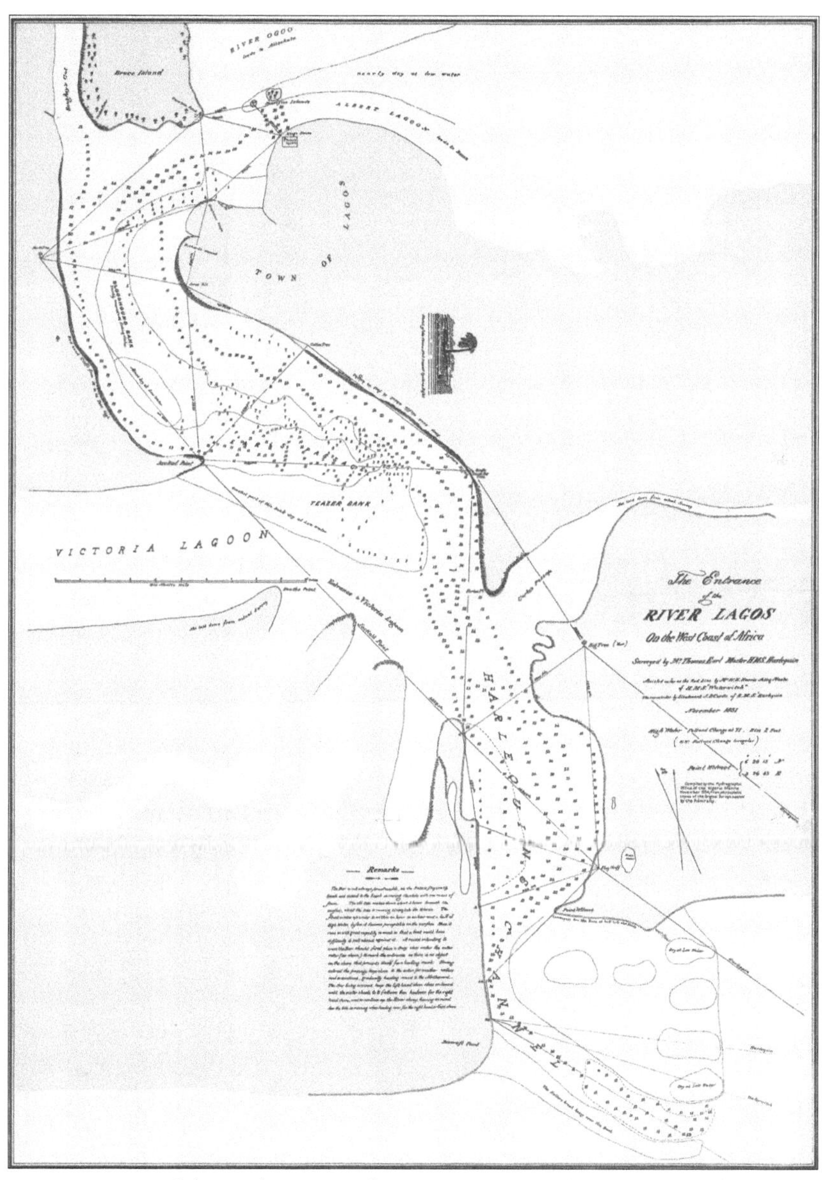

FIGURE 2.1. Detail from Thomas Earl, W. H. Harris, and J. D. Curtis, *The Entrance of the River Lagos,* November 1851, MR Nigeria S.127. Courtesy of the Royal Geographical Society, London.

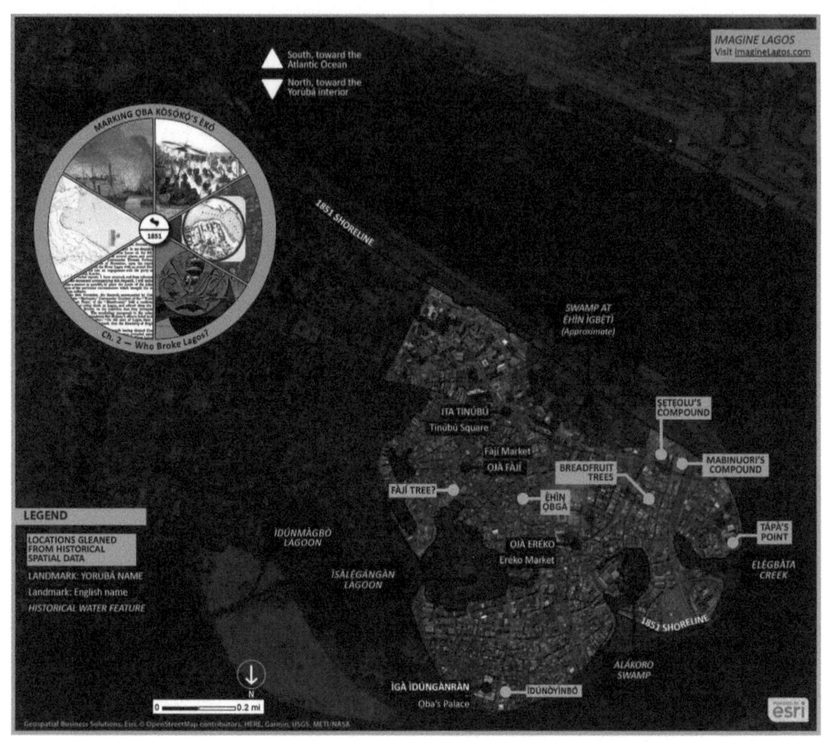

MAP 2.2. "Marking Ọba Kosọkọ's Eko." Base map courtesy of ESRI. Images in the cartouche are, clockwise from top right, as follows: (R1) detail of "James White Preaching the Gospel in Lagos," in James White, "Original Communications: New Prospects of Usefulness at Lagos; Extract from James White's Journal Dated January 10, 1852," *Church Missionary Intelligencer: A Monthly Journal of Missionary Information* 3 (June 1852): 124–25; (R2) Spyglass over Lake Street and Ẹlẹgbata Street; (R3) photograph of emblem at ọba's palace, Iga Idunganran, Lagos Island, 2018, photograph by author; (L3) detail from *Papers Relative to the Reduction of Lagos by Her Majesty's Forces on the West Coast of Africa,* vol. XXXVII.533 [1455] (London: Harrison and Son, 1852), 197, PP 1852, LIV.221, Great Britain, Parliamentary Papers (House of Commons Sessional Papers); (L2) historical map is Thomas Earl, W. H. Harris, and J. D. Curtis, *The Entrance of the River Lagos,* November 1851, MR Nigeria S.127, courtesy of the Royal Geographical Society, London; (L1) "Destruction of Lagos, on the Coast of Africa, by the British Squadron," *Illustrated London News,* March 13, 1852. Map by author.

bombardment in December 1851. The cartouche highlights the types of sources used: (clockwise from top right) an image of James White preaching; a Spyglass over Lake Street and Ẹlẹgbata Street; a photograph of the insignia on the Ẹnu Ọwa arch; an excerpt from *Papers Relative to the Reduction of Lagos;* an excerpt of a historical map; and an engraving of the bombardment of Lagos from the *Illustrated London News.*

The historical sites—such as the palace, barracoons, and markets—are plotted on a contemporary satellite image of Lagos as a way of reading the older sites in comparison to what the city is today. This interface between the contemporary city and Kosọkọ's Eko makes legible the historical markers of the "slave town" that have been erased or papered over in pursuit of a more "modern" Lagos. This map is in response to three spatial queries in the process of reading the city for evidence: If Eko was the name used by the inhabitants, how is it different from Lagos as a place? To what extent can we reconstruct Eko authentically, given that we mostly have data on targets destroyed by the Royal Navy? Many historical markers have been erased or papered over in pursuit of a more "modern" city; which are still legible, and why?

In map 2.2, I have marked significant sites in the city that are mentioned in the texts and maps. Some of them still exist today, such as the palace and Faji Market, so their locations can be mapped precisely. Other sites, such as Ṣẹtẹlu's and Mabinuori's compound, no longer exist, but were important to understanding the 1840s. These compounds represent the only two documented examples we have of the ara Eko who survived the bombardment. "Tapa's Point" is what the naval officers called the area controlled by Tapa in the Olowogbowo district, while "Idunoyinbo" is likely the area where the Portuguese lived while doing business with Ọba Kosọkọ. "Breadfruit" marks the site of St. Paul's Church, purported to be the site where enslaved people were sold and punished, while the barracoon in the southeast marked the only site we know of where enslaved people were held, possibly in the thousands.

A historian recently defined a city as an effect, or a consequence, of certain kinds of thoughts or ideas.[5] Cartographically, the Lagos we recognize today has a different orientation, making this map of the first understanding of old Lagos appear "upside down" in comparison. However, this map follows the logic of the priorities of Kosọkọ and his allies. Up to the 1850s, the Ẹhin Igbẹti area, the area on the Atlantic-facing shoreline, was considered the back of the city, while the palace was the front, linking Isalẹ Eko to the mainland and to other urban centers on the Bight. The bombardment and settlement of the British changed this political orientation, making the British infrastructure now the front of Lagos. Colonial maps of Lagos follow Cartesian logic, with the north at the top, which also has the effect of centering an imperial perspective and British settlement in the city on the Ẹhin Igbẹti by focusing on the Marina as the front of the city. Instead, map 2.2, in addition to recovering such historical markers of Kosọkọ's Eko, also considers what happens if this map is flipped and the 1850s perspective privileged. Instead of facing the Atlantic, we imagine a city with the palace facing the mainland, and the Ẹhin Igbẹti as the "back."

EKO: SUCCESSION, SLAVERY, AND AN EMPTIED CITY

Laurent Fourchard's insight that urban histories should be about politics is an apt starting point for reconstructing Kosọkọ's Eko, whose dense urban politics were structured with relation to the spatialities of (1) the urban circuits of the Bight of Benin, (2) the island and its lagoons, and (3) Eko's residential quarters.[6] My reconstruction of mid-nineteenth-century Eko's urbanism and its spatial politics focuses on kings and their entourages, not only because of their personification of power (and therefore place) but also because available sources provide too little insight about individual, ordinary Lagosians and other lagoon peoples between 1845 and 1851. Even local perspectives on Lagos's history preserved through origin stories, proverbs, and songs—many of which date back to the early seventeenth century—that are still remembered and sung in the city focus on these people. For instance, leading historical figures have their own *oriki,* or praise songs, in Yoruba that describe significant events during their lifetimes. Though many of these songs have not been precisely dated, they provide a distinctly indigenous interpretation of specific ideas and events. However, the meanings of each of these songs must also be deconstructed, as they are often a reflection of a single perspective, usually only of the dominant party. Few offer multiple perspectives on the same figure or event. So, for instance, Kosọkọ's defeat during the 1841 Ewe Koko War and the expulsion of him and his followers is remembered as a celebration of dispersal:

> *Mo le kan gb'ọna Egun,*
> *Mo le kan gb'ọna Epe,*
> *Af'eru baba tu yagba yagba loju omi.*[7]

This song calls attention to the significance of other urban centers on the Bight during succession crises, as well as the orientation of the city toward the lagoons and the mainland. Succession events through the 1840s and up to 1851 played out at the intersection of three specific spatial scales: the Bight, the island, and Eko's quarters. First, within the urban circuits of the Bight of Benin, Akitoye and Kosọkọ's competition meant that, depending on who had the upper hand in Lagos, they ricocheted between and drew upon support from towns like Badagry, Ẹpẹ, Ouidah, and Abẹokuta.[8] The influence of these urban circuits on ideas of succession also played a role in creating the succession crisis of the 1840s. Second, the changing relationship between land and lagoon (particularly because of the dangerous sandbar blocking easy access to the ocean, and the shallowness of the water on the southern side) affected not only the physical conditions and limits of the terrain but also how the ara Eko and others inhabited the space. The succession crisis of the 1840s altered the spatial politics of the island vis-à-vis the relationship between the ọba and Eko's chiefs. Third, the western portion of the island focuses on the areas where people lived, in quarters like Iga, Faji, Ọfin, and, beginning in 1841, Olowogbowo. The imbrication of these three spatial political valences was critical in the decade leading up to the British bombardment of 1851, as I explore below.

Lagos's kings were historically supported by four groups of chiefs who also occupied the island: the Akarigbere, the idẹjọ, the Ogalade, and the Aṣọgbọn. Historian Patrick Cole identified a hierarchy of their power based on their function and influence during Akitoye's reign between 1841 and 1845.[9] The Akarigbere were the royal chiefs and were the most important. Their head, the Eletu Odibo, functioned as the official kingmaker of Lagos society. Traditionally, the king was selected based on the consultation of the Ifa oracle. It is not clear if this position was hereditary, but Prince Kọtun, a Lagos historian and member of the ruling family, has identified six men who served in this capacity in the nineteenth century.[10] Next in line were the *idẹjọ,* or white cap chiefs, also known as the onifila funfun. Led by the Olumegbọn (the idẹjọ chief with claims to Idunmagbo and Aja), they claimed descent from the Ọlọfin, who was the original settler of Lagos Island. They were followed by the spiritual chiefs, or Ogalade, a group led by the Obanikoro. Finally, the Abagbọn, or war chiefs, were led by the Aṣọgbọn.[11] Kings typically had an extensive entourage made up of chiefs, wives, and slaves; none of these categories were mutually exclusive. Members of the household who were enslaved were known as ibiga and ilari;[12] however, there are several examples of enslaved men who later became important chiefs; for example, Kosọkọ's war chiefs Tapa, Ajẹniya, and Pẹlu. In the 1840s, the loyalties of these chiefs were split between Akitoye and Kosọkọ, and thus alliances on the island itself were important to not only manage its affairs but secure the power to do so. It is not always clear where each fell, but Tapa, Ajẹniya, and Pẹlu fought for Kosọkọ, while the Olumegbọn has been identified as loyal to Akitoye.

Much of the unrest in 1840s Lagos resulted from the struggle of two rival branches of the ruling family in Eko to control the island, which was part of a larger pattern of economic competition that animated the tension between cities on the Bight. The pattern of succession in Lagos was an odd mixture of Yoruba and Benin traditions because Benin had exerted influence over Lagos politics since the eighteenth century. Benin's political culture privileged primogeniture, enforcing the right of succession of first-born sons, while in many Yoruba traditions, the deceased's brother was supposed to inherit. One major source of the confusion was that, in Lagos, neither system had been strictly adhered to, as sons had inherited from fathers, and brothers from each other.[13] Historians Patrick Cole and Robert Smith have shown how the political rift between Kosọkọ and Akitoye dates back to the turn of the eighteenth century, when the former's father (Oṣinlokun) was passed over in favor of the latter's brother (Adele), changing the expected lines of inheritance from one of primogeniture to one where a more influential heir could be crowned king.[14]

Since at least 1836, Kosọkọ had been engaged in succession struggles over who would be king of the city but was passed over three times, and in each contest we see the significance of Eko's chiefs, especially the head of the Akarigbere, the Eletu Odibo. The first time was after the suicide of his brother Idewu Ojulari, who died with no heirs. In this instance, his right to succession was challenged by the Odibo, the town's kingmaker. The Odibo claimed the Ifa oracle had chosen Adele, another of Kosọkọ's uncles, but the truth of the matter was that Kosọkọ had made the Odibo his enemy by marrying a woman to whom he was betrothed.[15] Adele, a former king who was living in Badagry, was made ọba. When he died in 1837, Kosọkọ pressed his claim again, but Oluwọle, Adele's son, was chosen over him, again because of the workings of the Odibo. Then the Odibo accused Kosọkọ's sister Ọpọ Olu of witchcraft and infanticide, causing the king to banish her, even though she was found innocent of these charges. This was the catalyst for the Ewe Koko War in 1841, when Kosọkọ, Ọpọ Olu, and their followers declared war against the Odibo and the king.

During this war, Kosọkọ attempted to seize control of the Iga quarter—the seat of the ọba's power and the location of the palace—from Oluwọle. Kosọkọ and company were defeated, and fled across the lagoon to Badagry, after which Kosọkọ eventually settled at Ouidah. In 1841, after Oluwọle died in a mysterious gunpowder explosion,[16] Lagos was suddenly left without a leader; again, Kosọkọ and his uncle Akitoye were the major contenders.[17] Kosọkọ was still in Ouidah, and Akitoye was chosen and crowned in his absence. Akitoye remained king of Lagos until July 1845, when he was driven out by his nephew Kosọkọ during the Ogun Olomiro (Saltwater War), a twenty-one-day civil war between forces based in the Iga and Ereko quarters in Lagos.

Isalẹ Eko (marked as "Iga" on Glover's map) was the king's quarter and the location of his palace, the Iga Idunganran, and thus the epicenter of power in Lagos. Akitoye, described in contemporary letters as a "soft, charming and rather naïve creature," was eager for peace.[18] Against the advice of his council, he had invited Kosọkọ back to Lagos in 1841 and given him his own quarter, making him the *ọlọja* of Ereko, adjacent to the king's quarter.[19] Prince Kọtun argues that in creating this new ọlọja chieftaincy title in the Akarigbere class, Akitoye ensured he and his descendants would be excluded from ever becoming ọba, thus limiting Kosọkọ's ambitions.[20]

The Ereko quarter was located immediately south of Iga, and it was possibly his proximity to power that Kosọkọ considered rightly his that spurred his ambition. Even though Ereko is absent from contemporary maps (for instance, John Glover's 1859 map shows only the other four quarters), it is identified on an 1885 map as immediately south of the king's quarter. Its omission from the 1859 maps suggests that after Kosọkọ was defeated by Akitoye (and his British and Ẹgba allies) in 1851, the quarter receded in importance, or perhaps was taken over by new immigrants. In its place, Glover identifies Tapa's old quarter, Olowogbowo (rendered as "Oluwo"), as stretching from the southeast, south of Ọfin (rendered here as "Offee"), until it is adjacent to the southern end of the Iga quarter.[21]

The urban centers of the Bight afforded Kosọkọ the opportunity to cultivate new alliances. While exiled in Porto Novo and Ouidah, Kosọkọ became acquainted with the slave traders there, especially Francisco da Souza.[22] From there, he formed an alliance with the slave factions in Ouidah, Badagry, and Ẹpẹ.[23] Though Porto Novo and Ouidah were closer to the Atlantic Ocean, the increased efforts of the Preventative Squadron had decreased the profitability of the slave trade, and between the frequent blockades and seizure of ships, they were becoming less viable and secure as ports from which to ship enslaved people. Slave traders hoped to capitalize on their alliance with Kosọkọ by supporting him, in 1845, against Akitoye to regain control of Lagos, which, because of its labyrinthine lagoon system, was less accessible to the squadron's large ships and provided excellent cover for shipping slaves by canoe to the beach, where they could be loaded when the coast was clear.

While in power, Akitoye also cultivated relationships across the Bight. At Badagry, the missionaries were protected by Wawu, styled as the British chief, who also maintained a close alliance with Akitoye at Lagos.[24] When Akitoye became king of Lagos in 1841, he invited all the exiles from the Ewe Koko War back to Lagos. Pọsu, Ajẹniya, Kosọkọ, and Tapa returned to the city, despite the reservations of Akitoye's council. Kosọkọ, Akitoye's nephew,[25] was the most important exile. Before Kosọkọ's arrival in Lagos, the Odibo decamped to Badagry, arriving in 1841. He lived in Wawu's quarter, where he also courted the attention of the missionaries.[26] Soon after, Eko was split between supporters of Kosọkọ's and Akitoye's factions in Iga and Ereko.

Within the region, Akitoye was allied with factions at Abẹokuta, Badagry, Ajido, and Porto Novo, while Ouidah, Ẹpẹ, and Ijẹbu stood with Kosọkọ. The evidence suggests that Akitoye believed he and Kosọkọ could function as co-rulers, and that presenting him with his own quarter in the town would quell any ambitions to rule alone. By making him the ọlọja of Ereko, literally the owner of the market in Ereko, perhaps he would be satisfied with the full control of an entire quarter. He was wrong, and when he invited the Odibo back from Badagry to Lagos, Kosọkọ began to gather support to gain back the cap (read: crown) of Lagos.

The year 1845 was another turbulent one on the Bight of Benin because of competition between the Lagos ọba and the chiefs who supported and maintained his rule of the island, and the rulers of the towns on the lagoons. The largest of these towns, Abẹokuta, Lagos, Ouidah, and Badagry, were connected not only geographically by land, lagoons, and rivers but also commercially by their shared investment in the economies of the region. The profit from their shared markets in food, cloth, and other commodities was eclipsed by the income from the transatlantic trade. In each of these towns, well-entrenched Portuguese and Brazilian slave dealers conducted their trade with their local allies, and each town had a quarter that housed these traders.

Lagosians formed alliances with other lagoon cities in numerous ways, usually over economic competition and social grievances between prominent families. However, it was rare for an entire town to unanimously oppose or ally with another town, as there were always political, familial and cultural reasons to split one's alliances. There were multiple factions often delineated by quarters. In Badagry, for instance, a civil war erupted in June 1845 between the two factions that were loyal to the two warring princes at

Lagos. At Abẹokuta, Kosọkọ had allies in the transatlantic slave trade, even though Akitoye's familial connections ran deep there.

Profits from the transatlantic trade affected spatial and political relations. The power to create new quarters, distribute existing land (freely or at a price), and control movement within the island was increasingly in the hands of the rulers as the land-owning white cap chiefs saw a decrease in their powers and the king saw an increase in his own. Land was also tied to the appointment of new chiefs. In recording local testimonies about chieftaincy titles, Jonathan B. Wood noted that "those Chiefs made by the King were the more influential in political matters, but, at certain times, and for certain purposes, all of both parties were accustomed to meet together for consultation."[27] Tapa, Kosọkọ's head war chief, was one such example.

Rumors along the lagoons and rivers coursing through the urban circuits of the Bight often brought messages of gloom (which read differently to different constituents): of British cruisers patrolling the coast, waiting; of slave ships hidden in the shallow waters, lurking for a chance to resupply with human and other cargo; or even of war and weapons being cached. The Ogun Olomiro was preceded by a constant flow of rumors about the threat of Ọba Kosọkọ of Eko and his armed forces. In Badagry, in June 1845, the rainy season on this coast, one of these rumors bore fruit: Ọba Kosọkọ and his armed forces appeared on the beaches of Badagry, ready to attack.[28] At Badagry, thirty miles west of Lagos, rumors of attacks from Porto Novo[29] and Eko filled the air; from Dahomey, news of an attack to avenge a perceived slight; from Abẹokuta, news of allies ready to help; and finally, from Eko, stories of a return of the "Lagos people who have fled hither."[30] There was news that Ọba Kosọkọ and his uncle Akitoye were at war, that Kosọkọ "already has 2000 people in chains belonging to Aketoge [Akitoye]."[31]

The Odibo left Badagry on July 21, 1845, with a flotilla of around thirty canoes. He went east to Ajido, where his allies from Badagry were waiting. There, he assembled another twenty or thirty canoes, and started toward Lagos, on July 22, 1845, with a reported sixty or so canoes in his entourage.[32] In Badagry, the missionaries had settled on the eastern edge of the city and were protected by Wawu, the English chief there.[33] It was never clear that Badagry's entire population was in full support of the slave trade. Rev. Henry Townsend identified at least three different camps: those who wanted the trade to continue, those who wanted the English missionaries to stay, and those who were entirely opposed to the trade.[34]

In Akitoye's Eko, in July 1845, Kosọkọ was outnumbered, but not outgunned. John Losi, an early historian, mentions that Akitoye was ill-prepared for battle, even though he had more supporters in the town. "Women and children had been cut down, as well as men with arms, and almost the whole town was in ruins. . . . Of houses there were few undestroyed by the fire or other means during the fighting."[35] Kosọkọ and his supporters had set fire to the Iga, destroying all the houses in the quarter except the fireproof Iga itself. Despite this, he could not prevent the Eletu Odibo's party from landing on the island. In Rev. Charles A. Gollmer's estimation, Kosọkọ was more prepared and had acquired more ammunition and a "desperate set of followers."[36] Akitoye had more men, but reportedly lacked the fortitude for battle. Despite this uneven match, Gollmer believed Akitoye was defeated because he was betrayed by an unidentified chief (or headman), and by August 12, 1845, Kosọkọ had scored a decisive victory. The Eletu Odibo escaped but, according to some local traditions, was caught by Kosọkọ's men and beheaded.

By killing the Odibo and distributing the land to his war chiefs, Kosọkọ not only altered the politics, but also the arrangement of the city. Without the Odibo as the head of the Akarigbere, and further without the support of Akitoye as king, the Akarigbere saw the diminution of their power on the island, as seen in their disappearance from the sources until Akitoye was reinstalled in 1852. Without the Akarigbere class, Kosọkọ's war chiefs Tapa, Pọsu, and Ajẹniya saw an increase in their influence. Their promotion is also suggested in the city's layout in 1851. There are no white cap chiefs mentioned within Kosọkọ's entourage, and each section of the island was headed by these war chiefs, though it should have ordinarily been beneath the authority of the idẹjọ, remembered as the "original owners of the land."[37]

By the late 1890s, most men and women who commented on the precolonial organization of Lagos stressed the continuity of the power of the idẹjọ, with one arguing that the "land in Lagos was originally attached to the stools of the white-cap chiefs, who were in no way subordinate to the early Kings."[38] However, by seizing Lagos through violence in 1845 and by killing the kingmaker, Kosọkọ not only dramatically changed the politics of succession but also altered the structure of chieftaincies in the region. Before his rule, land was distributed by the free-born white cap chiefs. By taking the reins, he took the power of land ownership out of their hands and gave his formerly enslaved war chiefs—Tapa, Pọsu, and Ajẹniya—spatial control, inadvertently putting power over place into the hands of ex-slaves for the first time in the island's history.

Kosọkọ seized power, driving Akitoye to abandon his Iga and flee Lagos, first to Abẹokuta and then to Badagry, where he came into close contact with British agents through his developing relationship with the Christian missionaries ensconced there. He escaped, heading north through the Lagos Lagoon toward the Ogun River. Kosọkọ sent Ọṣodi Tapa to cut him off at the Agboyi Creek, a popular shortcut to the river. Back in the Iga quarter, according to the oriki, it took Kosọkọ three days to breach the guards who still held the Iga. He sent demands to Abẹokuta for Akitoye's head, but was ignored as Akitoye was protected by relatives on his mother's side, probably in Abẹokuta's Owu quarter.[39]

Meanwhile, at Abẹokuta, the king's death at the end of 1844 left the competing factions in disarray. Without a powerful leader to unite the town, the disparate quarters split according to their alliances around the lagoons. In April 1845, reports reached the missionaries at Badagry that the Ẹgba, the natives of Abẹokuta, had begun construction of a wall around their city.[40] By the middle of May, the wall and ditch had been completed.[41] Akitoye's presence in Abẹokuta proved deadly. Lagos was Abẹokuta's primary access to the Atlantic as the Ogun River connected both cities. Without Kosọkọ's support, they were cut off from the coast, and the only alternative route was to travel on land, a journey that was as long as it was dangerous. Akitoye was expelled from Abẹokuta, and immediately sought refuge in Badagry.

While in Abẹokuta, Akitoye was taken in by Sagbua, the new king, and housed in the Ake quarter near Ikija.[42] After only a few months, the increasing favor toward Kosọkọ meant Abẹokuta had also become unsafe for him, so he proceeded southward. By December, Akitoye had moved from Abẹokuta to Mowo,[43] a frontier town close to Ajido. A meeting was arranged with the chiefs of Badagry, and on December 23, 1845, the missionaries and chiefs joined him at Ajido. He expressed his desire to live in Badagry, and after much negotiation spearheaded by Mawu, he was invited to make his new home there.[44] Finally, on December 25, among much fanfare, he arrived in Badagry on horseback, making his entrance via the main marketplace.[45]

Clearly, Kosọkọ's reinforcements, ammunition, and provisions were coming from the western lagoon, maybe Ouidah, because as Gollmer remarked, had the people of Badagry "interfered" by "blockading the river," Kosọkọ would have lost. The effect of Badagry's apparent neutrality in the "rebellion at Lagos" was not clear at first, as in Gollmer's estimation they fared better under Akitoye than they would with Kosọkọ. Soon after, Kosọkọ sent them gifts to "flatter them" even though there were rumors he was plotting with his allies at Ado and Porto Novo to destroy Badagry.[46] Finally, the rumors that Kosọkọ was planning to continue the slave trade were confirmed when two slave ships appeared off the coast of Badagry a few months later, on September 7. The desire to control the port of Lagos was region wide; factions in cities, including Badagry, Abẹokuta, and Abomey, had their eye on the island or were threatened by its influence. Kosọkọ still maintained coastal and hinterland allies such as the Fon and Ewe to the east, and a host of Brazilian slave traders living in Lagos, Ouidah, and Porto Novo. But he also had a large number of enemies in his orbit. Among the most vocal were the Ẹgba chiefs at Abẹokuta and the Christian missionaries in Abẹokuta and Badagry. Unable to unseat him with their own armies, in 1851, the chiefs appealed to the British government to intervene at Lagos on their behalf: "I humbly and earnestly beseech you [Queen Victoria], therefore, to interfere on our behalf, to save our lives from the impending storm, and to prevent our being cut off as a nation, which you can easily do by overthrowing Kosọkọ and his *slave-town* [emphasis mine] Lagos, and reinstating Akitoye on his lawful throne there before Kosọkọ should be able to carry out his designs into execution, i.e. within the next two or three months."[47]

1845–51: THE REDUCTION OF A SLAVE TOWN AND THE ARCHIVE FOR RECONSTRUCTING IT

Will we survive the war this time? The women of Eko would likely have pondered their futures and that of their city—first in 1845, when a civil war, the Ogun Olomiro, forced them from their homes, and again in 1851, when British warships circled their island and city during the Ija Agidingbi. In December 1851, with the threat of Akitoye's British-supported invasion of the city looming, men stayed to build fortifications and rig the city's residential quarters with defensive cannons and guns, while thousands of women and children fled the violence through the lagoons toward other Bight of Benin urban centers like Badagry, Ẹpẹ, and Porto Novo. As they braved the cool harmattan winds,[48] they were seen transporting what they could fit in their canoes and small boats.[49] Of the estimated twenty-two thousand inhabitants of Eko, only a tenth or so remained on the island.[50] The others left their homes, markets, and compounds behind, places they had built (and indeed already rebuilt) by hand. Eko—as they called their city—would have to be made over again once the smoke cleared. Spikes and corpses would have to be removed from the creeks and lagoons so the fishermen could return to their boats, and the walls of their homes would have to be plastered to cover the damage from the shelling. These human and environmental tolls of the frequent, and often violent, upheavals have been overshadowed by political

FIGURE 2.2. "Shipping Slaves through the Surf, West-African Coast: A Cruiser Signalled in Sight," 1845, in "Slave-Trade Operations," *Church Missionary Intelligencer: A Monthly Journal of Missionary Information* 7 (November 1856): 241–55.

and economic histories that track the transition from the transatlantic slave trade to the "legitimate" commerce in the region.

The two civil wars, the Ogun Olomiro and the Ija Agidingbi, left the city in ruins, resulting in a lasting "reduction" of Lagos, in terms of both its physical urban landscape and also the silence around this period in the city's history. Both local oral narrative traditions and missionary records insist that this was an unremarkable, dark period better glossed over and forgotten. Oral narratives tell us that "nothing significant happened" during the reign of the "usurper at Lagos."[51] Missionaries, whom Kosọkọ would not allow in Lagos, were established on the coast at Badagry and thus unable to observe and comment upon happenings there. Kosọkọ's reign from 1845 to 1851 represents a rupture (though a less minor one than the British incursion) not only in the line of succession from Akitoye to one of his offspring but also seemingly in the urban history. Late-nineteenth-century colonial maps reinforce this image, showing that nothing was there that Europeans had not built. These ruins—and descriptions of them, including those descriptions used in planning the assaults that resulted in the ruins—as "privileged sites of reflection,"[52] provide evidence to reconstruct mid-nineteenth-century Eko, to evaluate its role within the lagoon community, and to examine its cultural, political, and economic reinvention during the end of Eko's participation in the slave trade. Reading Eko's ruins provides insights into the history of a region normally obscured by the lack of written and oral sources.

Slavery—in its domestic and transatlantic forms (see figure 2.2)—shaped and warped the urban fabric of the city, as testified by the accounts of the bombardment and Ọba Kosọkọ's removal. While attacking the city, British naval officers produced a written and visual record of this "slave town," tracking the most important historical actors while ever so briefly glossing over the people—Lagosians—whom they saw as collateral damage, thus allowing for a reconstruction of Eko as it existed at the end of 1851. These officers walked parts of the city, plotting it, before returning to their ships to attack it. After the smoke cleared in December 1851, some of them escorted Akitoye back to his palace, the Iga Idunganran, which had been mostly untouched by the shelling. They also drew at least three maps of Lagos Island, which supplement writings to provide vivid accounts of Lagos's urban form and the relationships between people and their environment.

On the morning of December 24, 1851, Lagos's six residential quarters—Iga (Isalẹ Eko), Faji, Ereko, Ọfin, Itọlọ, and Olowogbowo—once part of a bustling urban scene, were mostly deserted. The walls of houses and compounds were made over with weapons, and the streets were filled with Kosọkọ's armed men awaiting the remaining British ships to cross the sandbar and sail into the shallow Lagos Lagoon.[53] By midafternoon, the HMS *Bloodhound,* a British man-of-war anchored on the southern coast of Lagos Island, would have been entirely visible to the women and children who had not already fled the island.[54] That the ship was off the coast confirmed the rumors in town that the British were back and would avenge their defeat by Ọba Kosọkọ's forces in November. They had already burned down the barracoons on the beach, and the fire and smoke could be seen for miles around.[55]

The HMS *Bloodhound*—along with the HMS *Teazer, Harlequin, Volcano, Waterwitch,*[56] and others—was part of the British Preventative Squadron gathered on the "Lagos roads."[57] The squadron's stated goal was to eliminate the transatlantic slave trade in the Bight, either by negotiating a treaty with the king, Ọba Kosọkọ, or by force.[58] The treaty had failed, as Kosọkọ had rebuffed attempts to negotiate, but Akitoye, Kosọkọ's uncle and the previous ọba, was more amenable to abolishing the trade in enslaved people and was waiting in the wings to reclaim his throne, having gathered what troops he could muster from his allies in Badagry and Abẹokuta, and join the British in their attempt to remove Kosọkọ by force.

The Ogun Olomiro, in July and August 1845, is remembered in Lagos as the event that crystalized the divide between the two descendants of Ologun

Kutere, the fifth ruler of Lagos. This civil war between his son, Akitoye, and his grandson, Kosọkọ, ended with the partial destruction of the city and the introduction of the last regime that was economically and politically dependent on the profits of the transatlantic slave trade. After three weeks of fighting, when Akitoye and his men found themselves without supplies, surrounded by Kosọkọ's men, and forced to drink from the brackish lagoon that surrounded Lagos Island, they knew their town was lost. On August 12, 1845, they fled in their canoes and regrouped at Abẹokuta, ninety miles north of Lagos.[59] "At last," wrote one missionary, while recounting the story, "Kosọkọ was king of Èkó. But at what price, and of what a place?"[60]

The "price" had been the destruction of half the town, the death of thousands of its inhabitants, and the political and economic alienation of Abẹokuta and Badagry, cities that had supported Akitoye when he was king. Lagos, the "place," was a small city settled on the northwestern edge of Lagos Island. As ọba of this strategically located city-state, Kosọkọ controlled the land and sea routes on the Bight of Benin; Lagos was the only natural harbor in almost four hundred miles of the Atlantic coast. Abẹokuta, Ibadan, Ijẹbu, and Ikorodu, larger towns north of Lagos, depended on it as an outlet for their goods. Without an ally at Lagos, their economies were threatened, as they had to use the longer, more tortuous land route to the sea. Thus, Kosọkọ ruled the city for six and a half years on the support of profits from the trade in enslaved people.

In December 1851, between the twenty-fourth and the commencement of the bombardment on the twenty-sixth, British officers surveilled the city, from their gunboats or on the ground, recording information about the layout of the six residential quarters in Lagos and other important places, such as the beachfront, farms, and surrounding islands and water bodies. H. W. Bruce of the *Penelope,* Arthur P. Wilmot, the commander of the *Harlequin*—these officers were assisted in their mission by "men on the spot": other naval officers, consuls, and interpreters who walked Eko with them. Men like Captain Jones of the HMS *Bloodhound* and John Richards, an interpreter, first for Commander Frederick Forbes and then for Consul John Beecroft, had visited Lagos before the bombardment, making them slightly familiar with the layout of the city. They had conferences with the Brazilian slavers in Olowogbowo, perhaps in Idunoyinbo, and once even with Kosọkọ. Richards visited Lagos at least twice after the bombardment and had a clear view of different parts of the town. Though he left no record of his own observations, several accounts place him looking at, walking through, or speaking from different sites. The written accounts of Bruce and Wilmot, assisted by men like Jones and Richards, make it possible to envision the city and remap some of its spatial politics, even though it was in ruins by the end of the year.

In November 1851, to support the bombardment of the city, one naval officer produced a sketch of the city (see figure 2.1), which located the king's palace and other structures that were targets. This map explores in detail how sites in Lagos can be plotted using the letters in the *Papers Relative to the Reduction of Lagos,* and reports from missionaries, including Rev. Samuel Ajayi Crowther and Rev. Charles A. Gollmer.[61]

For three days, based on the intelligence gathered from previous visits, reconnaissance gathering, and surveillance from ships, the British bombarded Eko. The *Illustrated London News*' account of the bombardment of Lagos included an image that reproduced the view of the western bank of the lagoon from aboard a ship, showing the fortified city in the background and British boats landing in the foreground (see figure 2.3).[62] The pinnace (or small boat) and the paddle box from the HMS *Penelope* are depicted landing on Lagos Island to capture two of Kosọkọ's imported weapons: abandoned iron and brass guns. To the right are the spikes in the lagoon where it is around six feet deep. Another image from the article includes a Neapolitan gun, dragged to the ships by Kroomen and marines.

By about 3:45 p.m. on December 28, 1851, it was clear that Kosọkọ and his allies had fled the island. Captain Jones assembled a small group of men from the ships and some of Akitoye's chiefs to accompany him in canoes

DESTRUCTION OF LAGOS, ON THE WEST COAST OF AFRICA, BY THE BRITISH SQUADRON.

FIGURE 2.3. "Destruction of Lagos, on the West Coast of Lagos by the British Squadron," in "Destruction of Lagos," *Illustrated London News,* March 13, 1852.

to Eko and reinstate Akitoye as ọba of the city. Jones wrote his report about the treaty from the *Bloodhound,* which was then stationed at the north point of Lagos Island near where it had previously gone aground.[63] Ọba Akitoye signed the antislavery treaty in a large public ceremony at Iga Idunganran on December 31, 1851. A large crowd had gathered outside the palace at Ẹnu Ọwa; he hoisted a white flag with a diagonal red cross on it. Each article of the treaty was read aloud and then translated by John Richards; the crowd was said to have clapped their hands and clicked their thumbs after each one.[64]

Akitoye's return to his Iga marked the symbolic completion of the successful bombardment of Lagos, signifying the beginning of a new era (or maybe the continuation of a ruptured one). Akitoye's procession back to the Iga would have been a moment of bittersweet jubilation: he had his city back, but his authority was compromised by concessions allowing British officers and missionaries to settle on his island, and all around him were the ruins Lagosians would have to rebuild.[65] There are at least three significant pathways Akitoye and his party could have taken back to the palace. Walking around the town on Monday, December 29, they "found great destruction and havoc of property," as the bombardment had destroyed nearly half the town.[66] But what was his path back to power, and what do each of these potential paths signify in terms of the city's mid-nineteenth-century spatial politics? How do they reveal the slave trade's imprint on the urban fabric?

The shortest route would have had the British ships land directly in front of the palace. Adjacent to the palace were the destroyed homes of the three best-known Brazilian slave traders in the city: senhores Nobre, Lima, and Marcos. They likely lived west of the palace, around the area now known as Idunoyinbo Street, "where early Europeans made . . . their principal spot of settlement."[67] These men were also established in other parts of the city, especially on the beach on the southern edge of Kuramo Island. This path conjures the city's role in the slave trade and the significant population of formerly enslaved peoples who settled in Eko from the 1830s. Akitoye's path back to power depended on a variety of intermediaries and followers, made up of a handful of men who acted as interpreters for the English, and the troops from Badagry and Abẹokuta who had rallied for his cause. In their recent volume on intermediaries, Benjamin N. Lawrance, Emily Lynn Osborn, and Richard L. Roberts examine the pivotal role these men played in facilitating European expansion in Africa.[68] In the Bights of Benin and Biafra, Africans like John Richards were involved in several operations, and in the case of Lagos in the blockading of ports, the shelling of the island, and then in the aftermath of the attack as the smoke cleared. Richards was not so different from the Saro or other men, who later settled in Lagos. He had been enslaved after the destruction of Ijanna, his hometown. He endured slavery in Dahomey and was put on a Brazilian slave ship from which he was rescued before eventually becoming an interpreter.[69]

By virtue of being an interpreter for Forbes and Beecroft, Richards was present at many important moments during the attack on Lagos. He was at Beecroft's side assisting him when the *Bloodhound* was grounded in the shallow water west of Olowogbowo on the twenty-sixth and again the next day.[70] More significantly, he was part of a group of men who "escorted" Akitoye through to the city and to his house and installed him in his office. Captain Jones's reasoning for sending them with him was that "they who had joined him in adversity should have the honor of being his bodyguard."[71] Nearly 640 men had joined Akitoye to help in the attack on Lagos, and fewer than a hundred of them perished.

A second possible route for the procession would have started from the site of the old barracoons and would have provided another layer of significance to the British victory. On November 30, Commander Arthur P. Wilmot had ordered the barracoons and village destroyed.[72] Walking through the ruins of these structures would have cemented the idea that slavery was over in every part of the city, according to the terms of the treaty Akitoye had negotiated with Beecroft. But this site was too far to walk from the palace and would make any journey over land too challenging, as they would also have to make their way through swamps, over hills, and past charred bodies. But reflecting on the distance of the barracoons from the city demonstrates the nature of the practice of slavery in Lagos. These barracoons were at least ten miles from the center of the city but were close to many smaller villages and farms where people grew food.[73] The scale of this enterprise, when compared to the population of the city, is quite significant.

No matter which direction or itinerary Akitoye and his entourage took, he most likely passed through the spot where the triumphal arch was built in the 1860s. This junction, at the corner of Great Bridge Street and King Street, marked an important site in the city and was perhaps the historical location of Ẹnu Ọwa Square.

Akitoye's return not only signified the beginning of a new post–slave trade era for Lagos but the first imprint of the new imperial symbolic capital on the landscape, marking the early transition from Eko to Lagos. Although indigenous meanings persisted throughout the period, by 1851, the spaces of Lagos were invested with imperial ones too. As I use these sources to reconstruct precolonial Eko, I am reminded of the acts of erasure they contain. The reduction of Lagos was framed as dislodging a recalcitrant leader (Kosọkọ) to halt the transatlantic slave trade. The maps from the bombardment, however, record an additional insidious layer of ambition: the claiming of space and the territorial ambition that unfolded over the next decades. As Mark S. Monmonier has argued, "Military conquest and political revolution can precipitate a toponymy slaughter far more extensive than bureaucratic skirmishes over offensive place names."[74]

After 1851, naval officers projected themselves onto the space, renaming and recasting the city and island based on the names of their ships, shipmates, and captains. They renamed hills, slopes, points, and shoreline, mapping these spaces and lending a permanence to their presence. Some of these names remain in use in Lagos in 2023, decades after the bombardment, though others have faded from memory. Commodore Bruce had the former Ido Island renamed in his honor. Point Wilmot (no. 1 on map 0.5), where the barracoons and villages were located, was renamed after Commander Wilmot of the *Harlequin*. It was at this spot that he had beached the boats when it became apparent they were too heavy to pass the sandbar in November.[75]

Despite this imprint of empire, the sources provide information to remap Kosọkọ's Eko, and it is from these descriptions that the significance of the now-filled Ẹlẹgbata Creek emerges. See map 0.5 in "Dots and Lines on a Map: Note on Method," which, quoting Gollmer (see number 6), describes that most of the west—referring to Ọfin and Olowogbowo—has been destroyed, while there is a "large quarter that has not been destroyed at all."[76] Additional information on the town comes from Crowther, first from his memoirs published in 1837, when he was enslaved in Lagos (numbered 5 in map 0.5), and again when he visited Lagos in 1852 (numbered 2) after the bombardment. He talked of the transformation of the city, noting where a slave market had become a church.

The Ẹlẹgbata Creek, a key feature of the precolonial Eko shoreline, was covered up in the late nineteenth century, and the etymology of Ẹlẹgbata is not entirely clear, but is likely linked to Esu worship in Lagos. Despite the implied flatness of this two-dimensional map, these labels are not synchronic in the sense that they all emerged or were used in the same way historically. In fact, these labels have their own histories, and the differences in the deployment of Yoruba or English names have a specific history. Indeed, the fact that the basic alternatives are Yoruba/English confirms that the British and not the French or Belgians gained control of this space in the most "effective" way. Not every labeled point has a Yoruba equivalent, which suggests either that it was not recorded or remembered, or that a Yoruba equivalent was unnecessary.

1851, EKO: "A SCENE OF THE MOST PERFECT DESOLATION"

Contemporary reports estimated the number of the ara Eko in 1851 between twenty and twenty-two thousand, while the barracoons were thought to hold around five or six thousand people.[77] Considering the current cultural amnesia around transatlantic slavery as it was practiced in the city, this spatial distance reflects a historic willingness to separate transatlantic slavery from everyday life in Lagos. To the west of these barracoons was an entrance to the lagoons and creeks that stretched behind the coast to Badagry and beyond. The entrance to these creeks on the west side of the sandbank was renamed Point Bruce,[78] after H. W. Bruce's mission to destroy the barracoons. Using these creeks to transport their captives to another port would make it easier to evade the ships of the Preventative Squadron that could not navigate the shallow water.

The site of the barracoons was renamed Point Wilmot,[79] possibly because it was where Wilmot's ship grounded en route to destroying the "group of small huts and two stores belonging to the *senhores* Marcos and Nobre, Brazilians."[80] The barracoons had been emptied by the time the fire party arrived. What happened to the thousands of people who had been held in captivity there? There is no record of what happened, but we know that within a week of bombardment, a quarter of the population—approximately five thousand people—were back in Eko.

Despite the barracoons' distance from the center of Lagos, slavery was never a faraway idea inside the city. Many were enslaved in the city, in a system scholars have defined as domestic slavery.[81] In the heart of the Olowogbowo district, a specific site is remembered as the place of punishment, where enslaved people were regularly tied to a line of breadfruit trees. This site is the origin story for many layers of Lagos history, seen in the founding of St. Paul's Breadfruit in 1852, and the eponymous Breadfruit Lane discussed in the previous chapter.[82]

Before the British occupied Lagos, the Olowogbowo quarter was mostly a "steep bank descending into the lagoon,"[83] and the highest point was around seventeen feet above sea level. Most of the land on this side of the island, known as the "lagoon bank," was subject to the destructive forces of the "swift current of the ebb tide" and the force of the tropical rains.[84] Court records suggest this area was unoccupied before Ọsodi Tapa cleared the land and settled there with his servants.[85] This same source also indicates that Tapa's compound was likely in the parcel of land bound by Breadfruit and Broad Streets, between Bishop Street and Aṣọgbọn Street.[86] The name Olowogbowo possibly dates from before the bombardment, as it means "the owner takes their money back," alluding to the frequent capsizing of merchant canoes coming around the city from the Marina to Itọlọ.[87]

Given the effects of the bombardments and displacements, it is hard to understand who the actual owners of Lagos were. The sources identify fewer than five long-term occupants of the city by name, making it challenging to know who lived in Lagos and who survived the bombardment. In other words, who had a claim (outside of the chiefs) to land in places like Olowogbowo and Ọfin. Who owned the houses now reduced to rubble? These questions linger in the contemporary period. The consequences of this layering and displacement would echo into the twentieth century, as seen in records of court cases all over the city. A handful of court cases provide clues about previous occupiers of the city. If Akitoye had the time to wander here, they would have seen what was left of Tapa's compound, on the edge of Tapa's Point.[88] His house had taken the heaviest shelling and had been destroyed on the twenty-sixth or the twenty-seventh.[89] In the reports, this westernmost tip of Lagos Island was renamed "Bloodhound Point," echoing the site where the ship was grounded.[90]

Just as Kosọkọ had given Tapa the land, Tapa, in turn, distributed it to his own dependents. The area was filled with his family and other followers. Beneath the rubble were the homes of two men whose claims to the space have been preserved. One is Tom Mabinuori, who was himself formerly enslaved but had become free and created his own compound. Many years later, he identified himself as a "native of Lagos" in a land grant, asserting his belonging despite the frequent dislocations. The location of his compound was stated in an 1855 grant, in which he claimed to have occupied the site since at least the 1830s.[91] The second person was Ṣẹtẹolu, alias Gomez. He claimed to have acquired the land from Tapa at a time between the clearing of the area and the bombardment.[92] We know about this claim from a court case in 1916, where there was a dispute over the ownership of the foreshore, especially the steepest banks leading to the water. By tracing the lines of ownership back from W. B. McIver to Gerhard Johannsen and James Sandeman, we can locate the exact site, as seen in map 2.2.

There is very little documented about the two largest swamps close to the city, as the navy ships never ventured that far to the east in their mission to destroy the city. These swamps, Isalẹgangan and Idunmagbo, covered several acres of land and prevented the eastern expansion of the residential areas until much later in the 1890s. In fact, it was not until after the "slum clearance" schemes of the 1930s that these swamps were filled and the land occupied.[93]

All the evidence so far locates the inhabitants in this old quarter, which has since become identified with the Saro population. But before them, there was a sizable population loyal to Akitoye, Kosọkọ, or possibly both, depending on the fortunes of the day. A different source reports that Kosọkọ was given this area and was made a chief in the 1840s. Since the pattern in Lagos was usually that the land-owning chiefs led their own quarters, the evidence suggests that these two events likely happened at the same time, that his transformation of the unused land into Olowogbowo mirrored his rise through the palace ranks. By 1851, he was Kosọkọ's head war chief and was often described as his "prime minister," making him more important than Kosọkọ's other chiefs, Ajẹniya and Pẹlu, who, incidentally, had also been enslaved in their youth. The eastern half of the town survived the bombardment, which means, ironically, we have the least information about them. It was saved by Ẹlẹgbata Creek and the Alakoro Creek and Swamp, which had absorbed most of the fire from the rockets of the *Volcano* and *Teazer* on the twenty-seventh.[94]

Faji, Ajẹniya's district, was likely north of Ẹhin Igbẹti and was the site of a large farm and market. Faji is described as slightly isolated from the rest of the town, possibly because of the rather large swamps to the north and between it and Olowogbowo. It was also buffered by the defensive lines of the Ẹhin Igbẹti, which saved it from damage.

Throughout the day on December 28, Richards would have seen the people fleeing the city with all they could carry, rushing from Faji to Ido island.[95] There are at least two versions of the origin of the Faji quarter. The most popular versions attribute Faji as the sole quarter named after a woman, though evidence taken from one chief called Faji in the late nineteenth century offers a different account. The former traditions attribute the quarter to Fajimilọla, a priestess who migrated to the city from Imahin, in Egun, sometime between the late eighteenth and early nineteenth centuries, during Oṣinlokun's reign. After she helped the ọba, he made her a gift of land, east of the Iga Idunganran. Today, she is remembered in Lagos as a successful merchant, trading in goods and people. Her quarter, Oko Faji, probably began as a farm, as indicated by the name "oko." She established a market, the ọja Ita Faji, based on her profitable participation in the transatlantic trade.[96]

Christian missionaries, wishing to expand from their base in Freetown, saw opportunities east of them in the Bight of Benin. Agents from the Church Missionary Society saw the advantage in setting up a mission in Lagos, which was the ideal location for establishing their headquarters in the Bight, an island from where they could extend deeper into the vast and still densely populated Yoruba heartland. For the Christian missionaries, Lagos was the ideal location for establishing their headquarters in the Bight, an island from where they could extend deeper into the vast and still densely populated Yoruba heartland. Kosọkọ was their chief obstacle, as he was firmly opposed to Christian missionaries inhabiting his town.[97] Instead, the missionaries set up a mission in Badagry in 1844. From Badagry, they often petitioned the British officers of the Preventative Squadron, asking them to eliminate the menace at Lagos, ostensibly to end the transatlantic trade, but ultimately with an eye to expanding their mission there. Rev. Henry Townsend, one of the missionaries at Abẹokuta, explained the situation, writing, "The [British] Consul is in favour of Akitoye, the expelled rightful King of Lagos, who petitions him for Government to take Lagos, plant the British flag there and establish, him under it, and will make a treaty to abolish all Slave Trade and carry on only lawful traffic. I trust Government will take it up, for Lagos is certainly the focus of the Slave Trade, and will be a great acquisition, especially for [our] missionary operations."[98]

Akitoye, because of his alliances with the Christian missionaries and British naval officers in their desire to stop the slave trade, has been remembered as being against slavery and the slave trade; however, the nature of his alliances while in exile in the 1840s tells a different story. While exiled in Badagry, Akitoye was sure to deploy the rhetoric of antislavery in his letter to the consul:

> My humble prayer to you, Sir, the Representative of the English Government, who, it is well known, is ever ready and desirous to protect the defenceless, to obtain redress for the grievance of the injured, and to the triumphs of wickedness, is, that you would take Lagos under your protection, that you would plant the English flag there, and that you would re-establish me on my rightful throne at Lagos, and protect me under my flag: your help I promise to enter into a Treaty with England to abolish the slave Trade at Lagos, and to establish and carry on lawful trade, especially with the English merchants.[99]

His interactions with the missionaries, emancipated Africans and occasional British officers, naval and consular, would have demonstrated their candid interest in abolishing the transatlantic slave trade. Perhaps, more importantly, the activities of the Preventative Squadron, in capturing rogue slaving ships and blockading towns on the coast, would have demonstrated their military strength. In the Bight, as in other places, the British government complemented the actions of the Preventative Squadron by signing antislaving treaties with "native chiefs" along the coast. Eko was the last prosperous slave port on the so-called Slave Coast in the 1850s.

When approached by John Beecroft, the British consul of the Bights of Benin and Biafra, Kosọkọ refused to sign a treaty with the British. Beecroft then approached Guezo, king of Dahomey, who also refused to give up slavery or sign the treaty, claiming it was illogical to expect that his kingdom, eighty miles from the sea, was more of a threat than Lagos, which was only four miles from the sea. Additionally, the profits were too tempting, and far larger than those to be made from palm oil.[100] Beecroft then returned to Lagos. After a palaver on November 20, 1851, where Beecroft attempted to convince Kosọkọ one last time to sign a treaty and abolish slavery, Kosọkọ's response was that "the Friendship of England was not wanted."[101] Insulted at being spurned, Beecroft was instructed by the Foreign Office to approach Kosọkọ at Lagos a final time with a naval escort, and appeal to him again. Instead, on November 25, Beecroft attacked the island with a naval escort that included 306 officers and men aboard the HMS *Teazer* and *Bloodhound* (with twenty-one boats in tow).

Perhaps Kosọkọ acted wisely as he realized that the naval escort was rather peculiar, that those who "come in peace" usually do not bare their ammunition so openly. Perhaps no one was more surprised than Beecroft to see that Lagos was prepared for war. The *Teazer* and *Bloodhound* ran aground in the shallow lagoon and were hit with heavy fire from Kosọkọ's men. After realizing the futility of their efforts, Beecroft retreated in defeat. The losses on the Lagos side are unknown, while the British lost two officers and had fourteen men wounded. Once the news reached Britain, Beecroft was severely rebuked. He had attacked the island on his own initiative, exacerbating the sting from the failure of the mission. This failed November attack was captured by James George Philp, a British painter famous for his land and seascapes.[102] In his 1851 painting (now part of the Government Art Collection), he illustrates the heavy gunfire the British ships faced when they approached the island's southwestern shore. Of the island and city, not much is visible except the sandy coast, thickly ringed with trees that appear to be mangrove and palm. Here again, the city, in terms of its quarters, compounds, and markets, is invisible.

The British regrouped quickly, and planned a new series of attacks for December, this time with Akitoye and his men in tow. On the third day of fighting during the December attack, they hit Kosọkọ's main gunpowder magazine. By the next morning, December 28, the palace was deserted; Kosọkọ, and at least two thousand of his remaining troops, had fled to Ẹpẹ in sixty canoes.[103] The British had lost three officers and twelve men, with five officers and seventy men wounded. The Ẹgba sent troops to assist Akitoye and his British allies, but they did not reach the city in time. Similarly, the troops sent from Dahomey arrived too late to help Kosọkọ. News of Beecroft's eventual success in bombarding Lagos reached the London public by February 1852, but the press was ambivalent in its reports of the significance of the event. Lagos, on a tiny island in a far-flung territory seen as neck-deep in the slave trade (and not yet a colony), was of marginal interest, and reports were dubious about what was to be gained by sacrificing the lives of British officers to intervene in political squabbles. Despite the newspapers' indifference, the men on the spot were clear that Lagos was "the most important position that we [the British] held in Africa."[104]

~

After his ouster, Kosọkọ moved with Tapa and his other chiefs to Ẹpẹ. From here, they launched frequent attacks on Lagos. However, when they were allowed back to Lagos, instead of returning to Eko, Tapa and his dependents were settled in yet a new quarter on the east of the Lagos Island, which they called Ẹpẹtẹdo.

Ija Agidingbi is mostly forgotten in Lagos today, as its significance has been displaced by the British government's annexation of Lagos in August 1861. Nevertheless, the destruction of Lagos in 1851 marked the end of a pivotal portion of the island's and region's history. After his defeat in the Ija Agidingbi, Kosọkọ established himself comfortably at Ẹpẹ, thirty miles east of Lagos, on the eastern lagoons, to trade in slaves again. Unsatisfied with settling in this part of the lagoon, he continued to launch attacks on Lagos with his war canoes.[105] In the years following Akitoye's reinstallation in January 1852, the situation on the lagoons remained tense, exacerbated by Kosọkọ's unrelenting attacks on Lagos. Dissatisfied with his continued exile at his new base at Ẹpẹ, he and his chiefs and followers continued to plot against Akitoye and the British at Lagos. Pẹlu, one of Kosọkọ's war captains, petitioned Benjamin Campbell, the new British consul at Lagos, for permission to return, even though he had fought on Kosọkọ's side in 1851. Akitoye, true to form, permitted his request.[106]

Rumors of Kosọkọ's return plagued the city. With these rumors interrupting the "legitimate" trade and contributing to a heightened feeling of insecurity in the region, Consul Campbell adopted a new strategy in dealing with Kosọkọ and company.[107] On September 1, Campbell wrote to Kosọkọ asking him his intentions for Lagos, suggesting he give up his claims to the city. However, a few days after Kosọkọ's August attack, Ọba Akitoye died suddenly. Dosunmu, his eldest son, was installed as ọba of Lagos in September 1853. Choosing Dosunmu to replace Akitoye meant that Kosọkọ was passed over yet again, likely fueling his already deep-seated resentment. After the death of his longtime rival, Kosọkọ wrote to Campbell. "Today, we got to hear that King Aquitoy [*sic*] exists no more, and with more willingness we implore in the name of Her Majesty the Queen to return to our dear country, for which end we implore the protection of H.B.M. Consul and Admiral."[108]

Kosọkọ denied he had been trying to seize the city, and claimed that in fact he had only come to rescue his chiefs, Ajẹniya and Pọsu, and had been successful in this action. Distracted by other affairs, Campbell did not pursue the matter for a while.[109] However, hostilities continued between Kosọkọ and Lagos, with repercussions felt as far as Ijẹbu and Abẹokuta. By the end of 1853, Campbell decided to negotiate with Kosọkọ, calling a meeting between him and Dosunmu's men, on a small island on the lagoon between Lagos and Ẹpẹ. Dosunmu came with his white cap chiefs, and Kosọkọ arrived with Tapa, Ajẹniya, Pẹlu, and Lima, and by noon, both sides commenced their deliberations, with Campbell as intermediary. The meeting lasted four hours.[110] Both sides agreed to a peaceful solution, but the terms of this decision were hotly debated. Kosọkọ wanted to return to Lagos, presumably appealing to Dosunmu to follow in his father's footsteps when Akitoye invited Kosọkọ back to Lagos in 1841.

This time, Tapa added that Kosọkọ would return as a "private person," signaling that he had given up his territorial ambitions where Lagos was concerned. Whether Dosunmu agreed was unclear, but Campbell was firm in his response. He was not convinced by Kosọkọ's or Tapa's appeal and declared that it was clear to him that "two kings cannot live in one town." He added that "Lagos already had a king, Docemo [*sic*], who they intended to keep."[111]

Dosunmu was arguably the last indigenous ruler of Lagos who, with his white cap chiefs, had full rights to the land. His inability to maintain the security of the island meant the British had to come to his aid frequently after the expulsion of Kosọkọ in 1851, such as in the civil war that again threatened peace in Lagos and its environs. Like his father, Akitoye, before him, he was unable to successfully defend his kingdom, and in 1861 ceded the town to the British. Seen from the perspective of access to and control of space—on the island, on the lagoon, and on the mainland—Kosọkọ's, Akitoye's, and, eventually, Dosunmu's complex struggles for the city are more clearly understood.

Who broke Lagos, and to what effect? In 1852, the answer was Gollmer, the Christian missionary who had brought Akitoye and Beecroft together in Badagry. But asking "Who broke Lagos?" involves two more important directions. The first, answered here, is of course why, and the second is "Who rebuilt the city?" This query is explored in the next chapter, which discusses the early years of uneven local and European cooperation on the island. It looks at the development of the city during Akitoye's years, through the transition to Dosunmu's rule beginning in 1853.

3 ~ A New Eko?

At Water, Tinubu, and Ọdunlami

> There is one glaring fact, that must appear very strange, to everyone, that is, no one person, merchant nor missionary, English or Foreign, has been able to establish himself, except Mr. Gollmer, since the reduction of Lagos.
>
> —Vice Consul Louis Fraser to the Earl of Malmesbury

EVEN THOUGH British gunboats destroyed most of the city of Lagos in late December 1851, the city still holds traces of its past in its streets, squares, mangroves, and lagoons. To reconstruct these spaces and encounters in postbombardment Lagos, I walked the old Marina to get a sense of the changing orientation of the city discussed in chapter 2. In the consular decade, Water Street (as the Marina was known) was old Lagos's new front, and we walked to discover how this street still lodged memories from the early days of converting Lagos from a "slave town" to a new city where freedom was supposed to be a possibility for everyone. Yet, these early days were not without disputes that plagued the newcomers and relegated the ara Eko, or "native inhabitants" of Lagos, to the background.

In map 3.1 above, the Spyglass hovers above the intersection of the Marina and Tinubu Street and Ọdunlami Street. This junction was the site of one of the most troublesome debates over land and religion in postbombardment Lagos. West of Ọdunlami was the plot claimed by the agents of the Church Missionary Society (CMS) and is today the site of the Cathedral

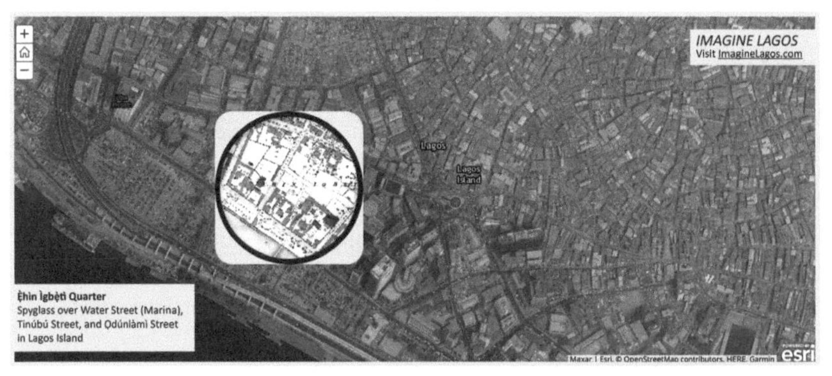

MAP 3.1. Spyglass over Water Street (Marina), Tinubu Street, and Ọdunlami Street in Lagos Island. Base map and Spyglass courtesy of ESRI. The historical map is a detail from *Plan of the Town of Lagos—West Africa (in 15 Sheets)*, 1891, 1 inch to 88 feet (scale), 63 × 90 cm, CO 700/Lagos 14, courtesy of the National Archives, Kew, UK. Map by author.

Church of Christ. West of this was the original site of the "iron coffin," which was the name given to the first British consulate built in Lagos. Both parties—the CMS and the British officers—had played a critical role in the bombardment (whether in setting the stage or manning the actual gunboats); and both parties tried to lay claim to "the best parts of town."[1]

Almost immediately after the city's partial destruction in 1851, questions of how to rebuild it emerged. Reassembling Eko from the rubble would never be easy. The city had to be remade, not only in terms of place but also in terms of *its* people. For some ara Eko, it was a strained homecoming, especially if they were returning with Akitoye. Would their homes still be there? Would Kosọkọ's people have altered the city, making it unrecognizable? Perhaps these questions did not matter everywhere in the same way, as residential districts like Olowogbowo and Itọlọ had been flattened.

The Old Marina Road was the only street I returned to numerous times during my research walks on Lagos Island. It was always a useful place to begin, especially when armed with the documents, letters, and maps that describe the transformation and division of the city. Whether I was at Olowogbowo to analyze the Breadfruit area and document the layering of slavery and freedom in the city, or in Tinubu Square to ponder why women were written out of archival records but starkly inscribed in the center of old Lagos, or even tracing out the histories of the three versions of Government House that were established and then moved east of Water Street in the nineteenth century, it was clear these events still lingered in the urban fabric of the city.[2] When the Marina area was known as Ẹhin Igbẹti—literally the back of the city—it had been a dumping ground for refuse, even corpses, and other abandoned things. The continued spatial incongruence of Marina Street demonstrated the tensions that developed as the city was resettled.

Lagos was remade to facilitate trade, to generate economic profit, and to remove the vestiges of the transatlantic slave trade. The city that emerged after the bombardment reflects these priorities spatially: in the settling of formerly enslaved Christian people, in the privileging of Christian missionary space, and in the sidelining of the local populations who were on the fringes of the new agenda. The records of the city and island focus most intently on the ways the new city emerged.

For many people, even those who had been enslaved in or from the city, the places in Lagos were real. They remembered where they had been enslaved, where they had been punished, and in some particularly bitter cases, where the homes of those who had enslaved them were and where they had been kept in captivity. For others, it was a return that was bittersweet. Despite the initial rush, some people delayed their return to the city. Rev. Samuel Ajayi Crowther had written about his enslavement in Lagos, including the places he was held and sold from; he did not join the initial rush. Others, like Gilbert Lawson, longed to confront the chief who had enslaved him and sold him into slavery.

These face-to-face encounters between formerly enslaved people and their captors, and between missionaries, administrators, and Lagosians, spurred letters, petitions, and maps that traced the nostalgia for home, land grabs through treaties, and quarrels over what many newcomers considered the "best parts of town." Eko and the ara Eko become visible in their measurements, debates, sketches, and disagreements over the distribution of land. Local elites, together with British officials and Christian missionaries as their allies, were at the forefront of reimagining and reinterpreting space in Lagos after the Olowogbowo and Iga quarters, the stronghold of Kosọkọ's power, were destroyed. Beginning in early 1852, newcomers to the city also began making crucial decisions about how the city would be rebuilt. This chapter analyzes the letters, journals, and maps of the new arrivals, reconstructing the space where they struggled to stake their claims.

Individual actors were swept up in the broader mechanisms of abolition, freedom, and emancipation in the Lagos lagoons. When the smoke cleared, the idea that a new Lagos was possible was heard all around the lagoons. Missionary letters spoke of "new prospects of usefulness" there, while merchants were eager to exploit the city's position as a "gateway to the interior."[3] An interior, they imagined, could switch quickly and seamlessly from exporting enslaved people to new commodities for the "legitimate" commerce: cotton, cash crops, and palm oil. At the same time, old residents, newcomers, and formerly enslaved men and women were returning to the city in droves. These categories are not necessarily mutually exclusive. Of all these returning people, the folks least documented were the local inhabitants of the city—both the ọmọ Eko and ara Eko. What had formerly been the domain of slave traders was now on the verge of being a new city. But how would its very recent history of slavery haunt the city?

The logic of the bombardment dictated that it had been done to displace slavery and the ọba that endorsed and profited from it. The letters flying around the coast had been full of talk of Ọba Kosọkọ as the "usurper king" who, with his partners, had led a "slave town." The new city would be Christian and legitimately commercial, though not necessarily in that order. Sometimes, these disputes had the tinge of implicit, in one case explicit, racial discrimination; one trader saw the southern edge of the island as a space for White residents, and them only, and said as much in his testimony.

Until 1852, all significant attempts to claim and reshape land on the island of Lagos were controlled by local elites—including the king and favored chiefs such as Tapa, who had been formerly enslaved but had been "owner" of the Olowogbowo district. They retained control of the distribution of land. By allowing the British to destroy the town, Akitoye effectively decoupled his political power from spatial control. This is further seen in his payment, through land, for their services, as he allowed the British consular officers and Christian missionaries to become entrenched in place and in many cases gave them free land.

The new map in this chapter, "A New City Springs from the Breadfruit Trees" (map 3.2), demonstrates the results of a walking cartography that tracks the tensions around acquisition and settlement and between two branches of imperialism: Christianity and civilization. Commerce played a role in these battles but took a back seat to the struggles between missionaries and administrator. The spatial material comes from an 1855 sketch, *Plan of Lagos and the Disputed Lands Claimed by the Church Missionary Society.*[4]

Benjamin Campbell's sketch of Water Street between Bishop and Apọngbọn highlights the tensions between European settlers. The Christian missions used their influence to gain land in Lagos. However, the men from the Preventative Squadron, who technically led the attack on former ọba Kosọkọ on behalf of Britain, also attempted to claim land for their sovereign. This intimate map is an example of the informal sketches produced in Lagos that draw attention to spatial practice and provide answers to the spatial queries around the rebuilding of Lagos and the resettlement of Liberated Africans from Sierra Leone. This map shows three plots of land on Back Street—later Broad Street—in the Ẹhin Igbẹti area on Lagos Island. Olowogbowo, described as the "quarter solely occupied by Sierra Leonian people," is visible on the left part of the map.[5] Several public lanes are labeled, and we see the roads that would eventually be called Breadfruit Lane, Bishop Street, and Apọngbọn Street, even though this sketch was created to show the tension between Campbell and the agents of CMS.

This rest of this chapter is in three parts, organized around the evidence that laid the ground rules for the reinvention of Lagos. First, it reconsiders a new geography on the island, made possible by the ruins of the city. Then it analyzes the process of rebuilding Lagos's quarters after the bombardment, as well as plotting the establishment and repurposing of new districts on the island. Finally, it looks at the tussles between two types of European power over space on the island: the Christian missionaries and the British consuls.

A WALKING CARTOGRAPHY: "A NEW CITY" SPRINGS FROM THE BREADFRUIT TREES

The walking cartography took root at the spot of the first missionary grant: at the intersection of Ọdunlami Street, Water Street, and Tinubu Street. The Christian missionaries did not respect the color line the administrators were trying to develop, preferring instead to live in both the new European corridor along the coast and in the indigenous quarters. Map 3.2 shows the pattern created by the five lots chosen by Rev. Charles A. Gollmer. The missionaries and merchants recognized the importance of the Atlantic-facing strip and wedged themselves into the most important parts of the city: on a former shrine, by the king's palace, in the destroyed quarter (Olowogbowo), over the home of a former slave trader, and almost in the center of town, facing the Atlantic.

The new map in this chapter, "A New City Springs from the Breadfruit Trees," demonstrates the results of a walking cartography that tracks the tensions between acquisition and settlement, and between Christianity and "civilization." Commerce played a role in these battles but in some ways took a back seat to the struggles between missionaries and administrators; however, both parties sought to establish bases from which to fill the economic, moral, and administrative vacuum in the wake of Kosọkọ's departure and the slave trade's termination.

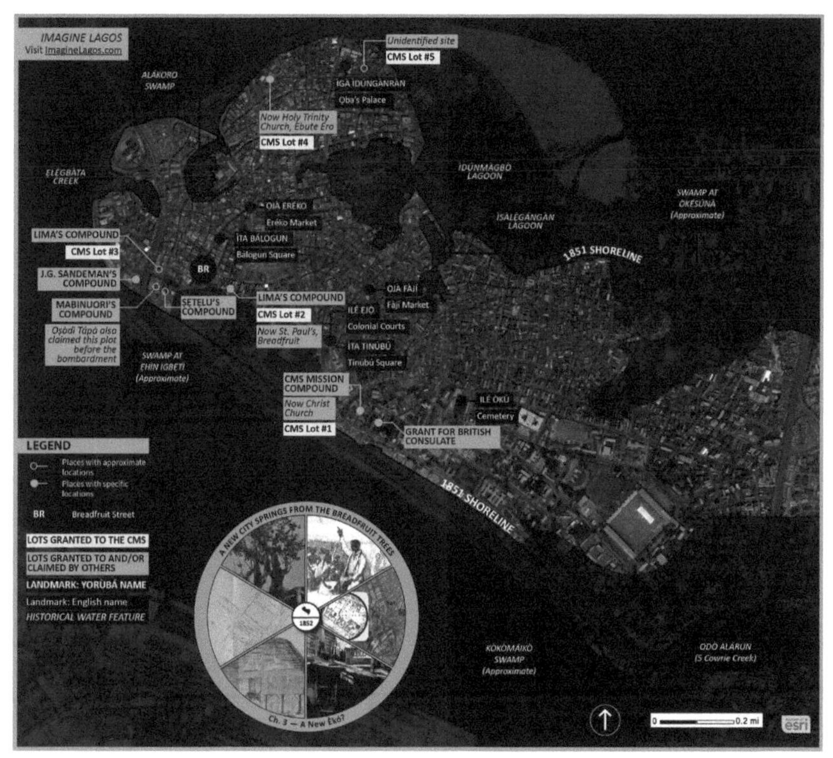

MAP 3.2. "A New City Springs from the Breadfruit Trees." Base map courtesy of ESRI. Images in the cartouche are, clockwise from top right, as follows: (R1) detail from "James White Preaching the Gospel in Lagos," in James White, "Original Communications: New Prospects of Usefulness at Lagos; Extract from James White's Journal Dated January 10, 1852," *Church Missionary Intelligencer: A Monthly Journal of Missionary Information* 3 (June 1852): 124–25; (R2) Spyglass over Water Street (Marina), Tinubu Street, and Ọdunlami Street; (R3) Breadfruit Street, Lagos Island, 2018, photograph by Lekan Adedeji; (L3) Church Missionary Society compound on Water Street, 1853, from inclosure in Slave Trade No. 6, FO 84/876, courtesy of the National Archives, Kew, UK; (L2) *Plan of Lagos and the Disputed Lands Claimed by the Church Missionary Society,* inclosure in Slave Trade No. 31, 1855, FO 84/976, The National Archives, Kew, UK; (L1) a photograph of the historic breadfruit trees in Olowogbowo, undated, in Archdeacon J. Olumide Lucas, *St. Paul's Church, Breadfruit Lagos: Lecture on the History of the Church (1852–1945)* (1946). Map by author.

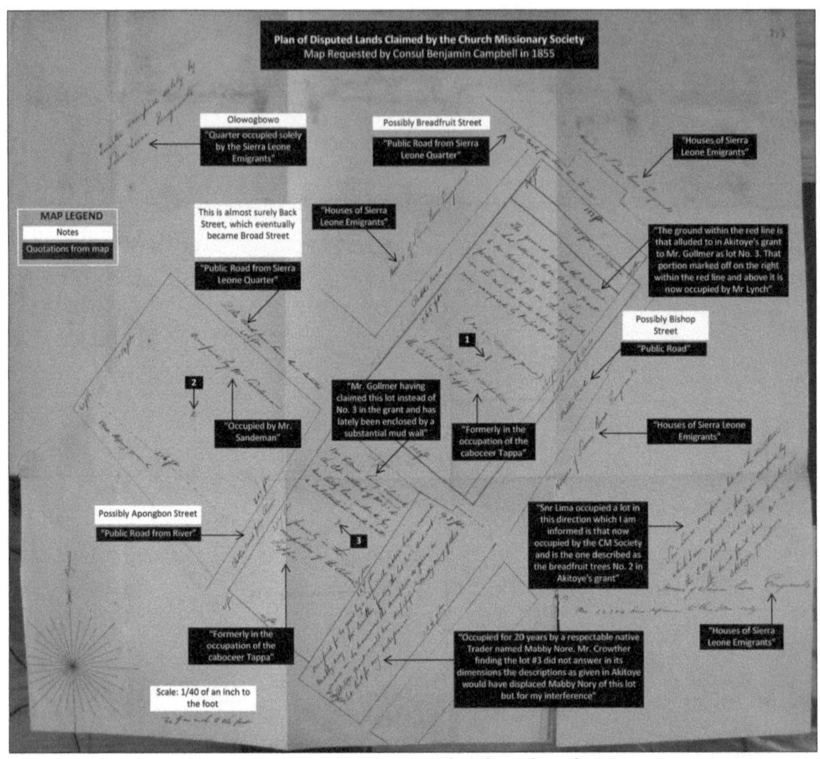

FIGURE 3.1. *Plan of Disputed Lands Claimed by the Church Missionary Society,* inclosure in Slave Trade No. 31, 1855, FO 84/976, The National Archives, Kew, UK. Annotations are by author.

Was Lagos a new city yet? This map plots the way Lagos was reconfigured in the early 1850s when Akitoye and his allies returned to the city from Badagry. Several new questions emerge about the identity of the city after the bombardment. How did the newcomers change the city? Did they all do it in the same way? Was Lagos a visibly different place once the missionaries and palm oil merchants had displaced the slave traders? A handful of ara Eko, new dwellers, and slave traders become visible in mapping space. Sites of punishment, enslavement, freedom, and reorientation dot the city.

Some of the spatial material in map 3.2 and elsewhere in the chapter comes from an 1855 intimate map, *Plan of Lagos and the Disputed Lands Claimed by the Church Missionary Society* (see figure 3.1). Created as evidence in Consul Campbell's complaint against the agents of the CMS, whom Campbell saw as taking more than their fair share of land in Lagos after the bombardment, the sketch is interesting on a few levels.

First, it demonstrates how the Saro (formerly enslaved people who migrated to Lagos; a Yoruba elision of "Sierra Leone") and Christian missionaries settled, claiming large sections of Olowogbowo. It also shows a handful of public and private streets in the 1850s, where no other evidence for them exists. Finally, and most importantly, it lists the former owners of the plots that are now occupied, and through this, we can locate the properties of slave traders Tapa and Lima in the old city.

Akitoye was said to have eagerly distributed some of the abandoned houses in Lagos. Evidence from testimony in a court case showed the retuned ọba "sent round his bell man saying that anyone who wanted possession of unoccupied land could take it"—and so they did.[6] The largest post-1852 population was the ara Eko, many of whom would have tried to reclaim their old homes, compounds, and markets. There was a smaller population who came to Lagos for the first time, many because they could be free there. Together, they outnumbered the other people, but they left very little physical trace of their presence, and even less documentation. We learn about their activities and ambitions from the records left by those who observed them, the smaller populations who, in acquiring land on Lagos Island, recorded their observations of and interactions with the ara Eko. The bulk of this information comes from the missionaries who obtained land all over Lagos Island, partially in the northwest, where the ara Eko found themselves increasingly corralled, and on the southern edge of the island facing the Atlantic Ocean, and within the sight of the British naval ships—the HMS *Harlequin, Teazer, Penelope,* and others—from whose decks they supervised the first few months of the rebuilding of Lagos. With the missionaries came a wave of "Sierra Leoneans," so called because they had either been born in Freetown or had been rescued by the Preventative Squadron and settled there, all the while being subject to various processes of "civilization" through instruction in English and encouragement to convert to Christianity.

Writing in 1852 to Henry Venn, secretary of the CMS, from the new Christian Missionary Mission House on the lagoon side of Lagos, Gollmer detailed the plots in Lagos that the CMS had acquired since the bombardment. Unfortunately, the map has not survived in the archival record, but the detailed nature of the grant he received from Akitoye still allows the approximate plotting of these locations. In this letter and others, there is evidence of a growing division between the populations. The Yoruba-speaking populations, though numerically dominant, became increasingly politically and economically marginalized. For the first time, we see the emergence of the idea of Eko's original settlement as "Old Town," a term that referred to the north and western quarters of the island, where the Black populations were found.

Though there is no claim to or evidence of the development of a (physically) coherent "Ne̩w Town" on the island, there is the emergence of a grouping of a handful of European missionaries and palm oil merchants in the Ẹhin Igbẹti, which was to become Water Street in the 1850s and 1860s, and finally the Marina, as it is still known in Lagos today. Despite the treaties and agreements negotiated with Ọba Akitoye following the bombardment, the initial process of acquiring land in Lagos took unexpected and tricky turns, and the disputes between the merchants and missionaries, and occasionally native Lagosians, form the backdrop for drawing a portrait of postbombardment Lagos.

After the bombardment, Gollmer and the CMS could settle in any part of the island. However, he settled in Ẹhin Igbẹti and reconstituted the Olowogbowo quarter with Liberated Africans from Badagry and Freetown. In doing so, the CMS was accused of attempting to displace longtime residents, who tried to reclaim their home after the bombardment.

By the time he received the plots of land in 1852, there was already a settlement of Brazilian returnees living in Lagos, but they had chosen to move east, outside the main core of the city. Brazilian Town emerged as an important site for Christianity and a distinctive area for architecture in Lagos as many returnees came back with masonry skills and often built in Portuguese style that reflected their experiences.

Olowogbowo also did not become filled with Sierra Leonean returnees by accident; Gollmer distributed the land specifically to them. Imagine the symbolic power of filling a residential quarter that used to be controlled by a slave trader and his accomplices with returnees who were formerly enslaved. All over Olowogbowo, we see the English names of the Liberated Africans who acquired land there: from the Coles to the Crowthers, to the Davieses, Savages, and Turners. Olowogbowo was reclaimed, and a new "west end" formed in the city of Lagos. At the core of the renewed district was an open square that once had enslaved people in barracoons. Those were stripped away and the reverend planted a row of breadfruit trees (see figure 3.2). This land became the beginning of Breadfruit Street, whose western end was marked by a new church, St. Paul's.

Of course, the ironies of this site were layered. Olowogbowo had been cleared by Oṣodi Tapa, a man described as Ọba Kosọkọ's prime minister. Oṣodi had been a war captive, and enslaved and sold, likely from the Bight of Benin. Unlike the Sierra Leoneans who occupied his district after his expulsion with Kosọkọ, he had been enslaved in the Iga Idunganran, most likely in the reign of Esinlokun, and through his own industriousness

FIGURE 3.2. The historic breadfruit trees in Olowogbowo. Archdeacon J. Olumide Lucas, *St. Paul's Church, Breadfruit Lagos: Lecture on the History of the Church (1852–1945)* (1946).

had been sent to Brazil.[7] An effective middleman, he could negotiate with the Brazilian traders, and eventually cast his lot with Kosọkọ rather than Akitoye. Ọṣodi is far from the only formerly enslaved person turned slave trader, but few, if any, of those in the historical record were displaced so literally by the people from whom they profited.

The first major distribution of land went to the CMS. At Gollmer's prompting, Akitoye gave the CMS five pieces of land in various parts of Lagos Island.

> Memoranda showeth,
>
> THAT Akítoyè, King of Lagos, with his Chiefs, has made over to the Rev. C. A. Gollmer, on behalf of the Church Missionary Society, the undermentioned pieces of land, for the purpose of erecting on them churches, schools and dwelling-houses for missionaries and native agents, whom that Society may employ at this station, Lagos.[8]

The largest plot, labeled "No. 1," measuring almost five and a half acres, was in Oko Faji, where Gollmer secured land for the construction of a church and premises for the CMS. With this land, he acquired a very valuable water frontage for the institution, cornering almost 130 yards of a prime lagoon in the part then known as the Ẹhin Igbẹti. The CMS grants are the only recorded grants in this period that are in proximity to the native population, while at the same time also being outside this area. Of the four remaining plots, Gollmer acquired a third of an acre near the ọba's market, as well as "Lima's Place," a one-acre property in Olowogbowo, close to where Tapa's house had stood before it was destroyed in December 1851 during the bombardment. The last two properties were a half-acre plot, west of Oko Faji, in the breadfruit trees area, and a third of an acre near Ebute Ero.[9]

Lot number 1 was in Ẹhin Igbẹti and was ideally situated as the first place travelers would see when entering the Lagos roads by ship or canoe. Marking this space for Christians, and facing the Atlantic, was a deliberate rewriting and reorientation of the history and meaning of the city. (See figure 0.6, Glover's 1859 map.) This location for the future Christ Church ensured the centrality of the institution. In fact, by 1859, it appeared as a graticule (intersection, or cross hair) of Glover's map of Lagos, with its coordinates of the "English Church" mapped out as an important anchor point for navigating to Lagos.

Lot number 2 was a plot that belonged to Brazilian slave trader Senhor Lima, who collaborated with Ọba Kosọkọ in the enslavement and export of people from Lagos. This lot was said to be the site of his large barracoons, and where Rev. Samuel Ajayi Crowther said enslaved people in Lagos were tied up and whipped. Gollmer planted a lane of breadfruit trees at this site, which eventually became the street name attached to the area, as well as the site of St. Paul's Church, Breadfruit.

Lot number 3 was another compound that belonged to Lima. Lot number 4 was in Ebute Ero, close to an important landing ground (*Ebute Ero* means "landing place of crowds"). It was also said to be a site for the shrine of Ota Omi, a local deity whom fishers in Lagos held in great regard. No evidence has emerged for the destruction of this shrine, but since many Lagos people were fishers, the significance of turning this site into a church cannot be overestimated.[10]

Lot number 5 was described as land on "the elevated part of Lagos behind the king's house, and not far from the market . . . called Ogogun."[11] This lot is difficult to identify despite the topographical and other landmarks mentioned. Ultimately, these plots were more than just land acquisitions for the church; in fact, they initiated a new regime that was only eclipsed or in competition with the British administration.

An announcement came early one August morning from messengers: "Akítoyè is no more: he died last night."[12] Akitoye was buried on February 20, 1854, and in Lagos this is noted as the date of the first recorded Adamuoriṣa festival in the city, known as Ẹyo Ọba Akitoye.[13] These proceedings were done to celebrate the lives of important Lagosians, and often began just before daybreak. The missionaries wrote that a "pretended corpse" made of cloth and covered with velvet was used to simulate Akitoye's body. Drummers, fifers, and singers performed to the right and left of the "body," and several objects decorated the scene. A sword was next to the body and a crown was placed upon it. Blue, green, and purple velvet adorned the walls, and pictures, glasses, and a clock were added. Onlookers and strangers who stood in front drank rum from "white square bottles," while the king's relatives sat next to the body. Cowrie shells covered the floor.[14]

The drumming and singing continued for days, and the followers of several deities "came by in turns to rejoice with the king." The missionaries noted that the proceedings carried on in the traditional way, except now there was no human sacrifice involved. Once all the proceedings were concluded, Dosunmu ended his mourning, shaved part of his head, and displayed himself in "a beautiful attire." His people followed suit, except that the free were distinguished from the enslaved, who had to shave their entire heads. Now that Akitoye was an ancestor, "numbered among the dead, is deified," all the people of Lagos (except the Christians) offered their prayers to him.[15]

OF THRONES AND TREATIES

The British naval officers who reinstalled Ọba Akitoye set the initial conditions for resettlement on the island. The scene aboard the HMS *Penelope* was a prelude to Ọba Akitoye's installation on the throne in Lagos and offered a template for the new organization of power on the island. Early on the Thursday morning of January 1, 1852, Akitoye and a few of his chiefs were on board the *Penelope,* where they had been invited by John Beecroft, British consul of the Bights of Benin and Biafra, and Commodore Henry William Bruce, the commander-in-chief of the British ships and vessels on the west coast of Africa.[16] While most of the population slowly trickled back into the mostly destroyed city, Akitoye and his newly enfranchised retinue of wives, chiefs, and children agreed to a list of treaty conditions that ceded his control of Lagos Island.[17]

Treaties like this, which were negotiated and enforced all over the West African coast in the nineteenth century, have typically been analyzed in terms of their political content, but in this chapter, I read them for their impact on the spatial organization of Lagos.[18] Embedded in these documents are the new conditions for the occupation of the city, and in this case, the "Engagement with the King and Chiefs of Lagos" outlined several conditions for the navy's continued defense of Akitoye from Kosọkọ's continued attacks.[19]

In signing the treaty, Akitoye and his chiefs agreed to nine articles in exchange for regaining his throne. The first five articles addressed the question of the slave trade on the island and in the lagoons, expressly banned the continuation of any activities related to it, and called for the expulsion of slave traders and the destruction of any infrastructure on the island that was used specifically for the trade. Articles VI, VII, and VIII established principles for the city's economy, especially in terms of welcoming European merchants to the palm oil and cotton trade without restriction, and for the city's religious future.[20]

While the first five articles clearly established the ascendent prerogative of the British in Lagos, establishing the right of British agents to act against the slave trade if the king or his chiefs failed or were unable to do so, they also constituted the basis for those agents and British missionaries and merchants to set the agenda for the city's spatial priorities. The second article made an explicit connection between slavery, the slave trade, and infrastructure on the island: "No houses, or stores or buildings of any kind whatever shall be erected for the purpose of the slave trade within the territory of the King and Chiefs of Lagos."[21] This justified the past destruction of barracoons during the second bombardment in December 1851 and forestalled their repair or the rebuilding of new ones.

Article IV stipulated that any "barracoons or buildings exclusively used in the Slave Trade, shall be forthwith destroyed." Additionally, Article V ensured that any existing "houses, stores, or buildings hitherto employed as slave-factories, if not converted to lawful purposes within three months of the conclusion on this Engagement, are to be destroyed."[22] While the navy, as representatives of the Crown, set the treaty terms that were critical to Lagos's new spatial arrangements in the 1850s, it was the missionaries who really set the agenda based on those terms.

The 1852 treaty was foundational to missionary activity in Lagos. Article VIII, the longest of the articles, allowed Christian missionaries to proselytize unmolested. It said, "Complete protection shall be afforded to Missionaries or Ministers of the Gospel, of whatever nation or country, following the vocation of spreading the knowledge and doctrines of Christianity, and extending the benefits of civilization within the territory of the King and Chiefs of Lagos."[23]

The ọba was to ensure that the missionaries were not "hindered or molested in their endeavors to teach the doctrines of Christianity to all persons willing and desirous to be taught."[24] Finally, the treaty addressed the specific way in which the island would be transformed spatially, including conditions for the distribution of land to Akitoye's allies, promising that "encouragement shall be given to such missionaries or ministers in the pursuits of industry, in building houses as their residence, and schools and chapels."[25] It is here that we see the basis for the distribution of land in Lagos Island.

This same treaty also made provision for building a Christian cemetery, noting, "The King and Chiefs of Lagos further agree to set apart a piece of land, within a convenient distance of the principal towns, to be used as a burial-ground for Christian persons."[26] It was with this mandate in hand that Gollmer, head of the CMS, set about acquiring land in Lagos and became the first European to claim acres in Olowogbowo, Faji, and Ọfin.

The treaty of January 1, 1852, also provided protection for British merchants in Lagos: "The subjects of the Queen of England may always trade freely with the people of Lagos in every article they wish to buy and sell in all the places, and ports, and rivers within the territories of the King and Chiefs of Lagos, and throughout the whole of their dominions; and the King and

Chiefs of Lagos pledge themselves to show no favour and give no privilege to the ships and traders of other countries which they do not show to those of England."[27] Barely two months after his return to Lagos, Akitoye and his chiefs signed yet another treaty in his palace, this time negotiated by the European merchants that had gathered at Lagos.[28] This treaty was far lengthier, with fourteen original articles and two additions governing how the merchants would be protected while monopolizing the brisk palm oil trade developing in Ẹhin Igbẹti. The first two articles established the import and export tax on goods in Lagos and that merchants would be free from maltreatment.[29]

Article III was the most explicit in terms of merchants and space. It stated, "That the King shall allow the traders to erect casks and storehouse on the east point entering the river, and as far as the passage which divides Lagos from it."[30] He also agreed to build a "Custom House" at this point, and choose a man or office to supervise the landing and shipping of goods. Further, he was to ensure that no Lagosians could build houses or commercial establishments there, thus maintaining a European monopoly on the southern edge. Ostensibly, this part of the treaty was included to "prevent the possibility of fire or theft."[31]

Perhaps the most significant point and the greatest challenge to the ọba's authority was found in article X, which specified, "The merchants settling at Lagos shall be allowed to choose their place of residence."[32] Almost all of them chose to live near or above their stores on the lagoon. The fourteenth article maintained that a copy of the agreement be held in the Custom House and be shown to anyone entering Lagos for trade. However, this audience was extended within the terms of the second additional article, which declared that the terms of the agreement "be publicly made known to the inhabitants of Lagos, by the appropriate person."[33]

The political and economic ramifications of the bombardments were felt for years after the island was rebuilt. In fact, it was only after the initial crisis over the partitioning of land between the merchants and the missionaries that people began to reflect openly on a third option: the decisive role that the Christian missionaries and their Ẹgba adherents at Freetown, Abẹokuta, and Badagry played in the reorienting of lagoon politics in the early 1850s.

THE BEST PARTS OF TOWN

The division of land in postbombardment Lagos was not without incident. In fact, disputes over property lines and the reclamation of space would punctuate much of the period between the bombardment in December 1851 and the final annexation of the island in August 1861. Even though these disputes replayed themselves on different scales all over the island, only the cases outside of Iga, the king's quarter, have been preserved in the documentary record. This was a time of acquisition and reinvention in Lagos, and the merchants, former residents, and formerly enslaved people who flocked to Lagos took full advantage of this, claiming land, anchoring themselves in the space of the new city, and transitioning to a prominent role in the palm oil business.

Christian missionaries, led by Gollmer, were able to negotiate with the king quickly and leveraged their role in the displacement of Kosọkọ to claim their choice of land. Merchants like James Sandeman found themselves sidelined, despite negotiating their own treaty with the king. In 1853, the Foreign Office appointed a new consul, Benjamin Campbell, to replace Vice Consul Louis Fraser in Lagos. Campbell's inability to get the land of his choice to build a consulate on the island's southern edge sparked a fierce debate, the ramifications of which were felt as far away as London, as their correspondence was forwarded to and from the headquarters of the CMS in London and to the Foreign Office.

One afternoon, in late March 1855, J. P. Boyle and S. B. Williams,[34] two Sierra Leonean men employed by Campbell, were measuring the plots of land owned by the CMS on the southern shore of the island in the district of Oko Faji. The new consul had spent his first three weeks in the CMS compound when he first moved to Lagos, and thus had his heart set on acquiring land in the same area. They were *remeasuring* the land as the consul had sent them to ensure the area marked off by the CMS's walls did in fact correspond to the grant they had received from Ọba Akitoye. Campbell was in search of "a moderate-sized suitable piece of the vacant ground adjoining the Church Missionary inclosure, for the use of Her Majesty, her heirs and successors."[35]

When Boyle and Williams were satisfied with the measurements, they reported back to Campbell, who then requested land from Dosunmu. In order to secure the land, Campbell returned to the site with the king. On reaching the area, they were confronted by Reverend Crowther, the new head of the CMS, who had likely been alerted to all the activity around the compound. Crowther immediately challenged Campbell, arguing that he had no right to any land in the region, and that, in fact, the CMS owned this entire district. Campbell's protests that the land the CMS measured corresponded to the original grant had no impact on Crowther. Even Ọba Dosunmu had no say as Crowther referenced an agreement made between Dosunmu's father and Gollmer, the former head of the CMS. Feeling defeated, Campbell erected a temporary shed to protect the building materials for the future consulate, but was determined to pursue the matter further. In his letter to the Foreign Office, he described the incident as a "continuation of the systematic annoyance exercised towards me by Mr. Gollmer and his colleagues."[36]

Of course, Gollmer denied any knowledge of a "long standing dispute" between him and any other men in Lagos.[37] The agents of the CMS and Gollmer specifically claimed that Campbell had "unjustly taken possession of a piece of land belonging to the Society whereon to erect a Consular residence."[38] Soon after, Campbell was called upon to defend his actions, as the CMS agents in London had written to his superiors that he had been persecuting the missionaries in Lagos. In assembling his defense, later published as *Correspondence Relative to the Dispute between Consul Campbell and the Agents of the Church Missionary Society at Lagos,* Campbell provided evidence from numerous native and European merchants who had similar encounters with the missionaries, not only in Faji where Sandeman had been bullied out of his own land, but in Olowogbowo as well, where natives like Tom Mabinuori had also been in danger of losing possession of his compound.

The land dispute case was divided between four specific parties. First were the European missionaries from the CMS, whose thirst and desire for land as the "firstcomers" meant they had their share of the choicest land they wanted, beginning on January 1, 1852. Next were the merchants who were interested in the palm oil trade. We see here the actions of German and British interests, as they went about trying to get land from the king, but they did not all have equal access and were not all treated the same. There were also the Brazilian and Sierra Leonean immigrants, who were given land. And, finally, were the British consuls. Akitoye was reinstalled by three parties: the British men-of-war, the Christian missionaries, and their shared Ẹgba allies. According to the documentary record, only two of these groups received land.

Despite Akitoye's agreements to support the merchants interested in entering the palm oil trade, part of the problem was the scarcity of suitable available land. Much of the land was swampy and covered with thick mangroves, and the livable areas were already occupied by the Lagosians returning to the city. "There are not more than four or five good spots, suitable for merchants, in all of Lagos," wrote Fraser (consul of the Bight of Benin) in March 1853.[39] "I am fully convinced," he continued, "that Mr. Gollmer has no use whatever for river frontage, beyond that occupied by his private residence, on Lower Fadge."[40] Sandeman's account was quite similar. In describing his experience in 1854, he wrote,

> I went to Lagos to look after a piece of ground to erect a house on and could not get a spot, every place, apparently belonging to Mr. Gollmer. All the Oloboo [Olowogbowo] district, excepting the spot I at present reside on, he had given to his Sierra Leone followers, and he claimed all the land from his own house to the Five Cowrie Creek. He sent me word by my messenger, "If I thought he was a black man, to take a piece of ground and not the river-frontage?" referring to that piece of ground formerly Tappa's [*sic*] and now the proposed cotton store.[41]

Part of the problem stemmed from the fact that the merchants had almost no political influence over Ọba Akitoye. Gollmer and the CMS were among his closest allies while he was exiled in Badagry and were now exercising their influence by acquiring whatever parts of the island they felt were suitable for their purposes. In addition, their closeness to the Sierra Leonean populations that came with them, mostly from Freetown, also secured their own access to the choicest parts of Lagos that could be reclaimed. Other immigrants, for instance those from Brazil and Cuba, did not have the same ties to the CMS. In 1853, it was reported that 130 Brazilian families residing in Lagos were fending for themselves and living "without protection" on the island. In December 1853, the consul took them under his wing, and by 1855, we see land grants to the Brazilian population, beginning with a grant of land in the "field at Oko Faji" to Clara Maria Lisboa, an immigrant from Bahia.[42]

Whether an exaggeration or not, Gollmer's influence over Dosunmu caused many merchants to chafe. In his complaint to Campbell, Sandeman added, "I was perfectly aware that Mr. Gollmer was de facto King of Lagos; but I think it was carrying it a bit too far to expect me to ask him for permission to occupy my own land."[43] Longtime residents of Lagos were not always secure in their ownership of land, as demonstrated in Mabinuori's clash with the agents of the CMS over his plot in Olowogbowo.[44] Campbell was opposed to what he described as Gollmer's "chopping and changing of one piece of ground for another."[45] He wrote to the Earl of Clarendon, explaining that Crowther had been unaware of the inaccuracies in Gollmer's measuring of the CMS plots, and when he found out that the CMS's claims infringed on Mabinuori's land, he immediately dropped the claim to his Olowogbowo land.[46]

The spatial struggles in postbombardment Lagos were defined by the conflicts among Christian missionaries, British officials, and merchants: it was never clear who would get the upper hand even though British ships had secured the town through violent means. Between these groups were the overlooked Lagosians, and as their city was increasingly divided up, they were relegated to the edges of commercial and political power.

As firstcomers, the agents of the CMS were quick to redefine space, even changing the spatial orientation of the town. The maps of the city show how their occupation of the central part of the waterfront, right in the center of the new Atlantic-facing commerce, was designed to secure Lagos as a Christian space, and by extension, to continue to convert in the interior. Recognizing the Atlantic-facing shoreline as the direction of the future, they were quick to secure spaces for themselves while also maintaining places in the interior to ensure their proximity to the king. By attaching themselves to the missionaries physically, the British consuls recognized this potential, but quickly noted the rivalries that emerged from trying to secure the "best parts of town."[47]

4 ~ Recovering Lost Ground in Old Lagos

At Joseph and Campbell

> Still, the Christian cemetery or Burial Ground, where the earth opens up her bosom impartially to receive both the PRINCE and the BEGGAR on a common level is a memorial and a record. It is not, when consecrated, a mere field in which the dead are stowed away unknown; no. It is a touching and beautiful history, written in family burial plots, in mounded graves, in sculptured and inscribed monuments. Such a burial ground tells the story of the past—not of its institutions nor of its wars or of its ideas, but of its individual lives,—of the men and women and children of the past and of its household. The Christian cemetery is silent but eloquent it is, but it is unique. We can find no parallel to such history elsewhere; there are no records in all the wide world over, in which we can discover so much that is suggestive, so much that is thought-provoking, pathetic and impressive.
>
> —Herbert Macaulay

IN DEATH, just as in life, Lagosians found their city to be a crowded place. The Spyglass in map 4.1 hovers over the site of the first public cemetery in Lagos, located at the junction of four quarters: the Brazilian quarter, Houssa Lines, Oke Itẹ, and Oko Faji. Herbert Macaulay's 1940 words in the epigraph remind us of a cemetery's potential for storing history, especially in a city like old Lagos where so many ordinary people are written out of the traditional archival texts and maps available.

MAP 4.1. Spyglass over Joseph Street and Campbell Street in Lagos Island. Base map and Spyglass courtesy of ESRI. The historical map is a detail from *Plan of the Town of Lagos—West Africa (in 15 Sheets)*, 1891, 1 inch to 88 feet (scale), 63 × 90 cm, CO 700/Lagos 14, courtesy of the National Archives, Kew, UK. Map by author.

This crowding came about partly from the influx of migrants to the city, beginning after the British Navy's bombardments of Lagos in December 1851. In January 1852, the British administrators requested land from the ọba to be set aside as a burial ground for Lagosians. The land they chose was originally outside the city's residential quarters, and in choosing this space, imagined then as far away in "the Fields" at Oko Faji, the administrators thought the cemetery would be safer and more sanitary.[1] As reported in chapter 1, this idea was met with the "deepest horror" by most Lagosians, who before this preferred to bury their dead at home.[2] Eventually, the cemetery became popular—too popular in fact—and by 1872, it was close to overflowing. The city was expanding rapidly eastward into and beyond Oko Faji, and already the streets had stretched beyond this space.

These spatial references—such as to "the fields at Oko Faji," the cemetery, and the "land at the water side called Ebete [*sic*]"—marked out the rapid expansion in old Lagos and are no longer legible on contemporary maps of the city. Inspired by the locations, both known and unknown, referenced in Ọba Dosunmu's seventy-six land grants—such as the cemetery, the "fields at Oke Popo," and Oko Faji—I investigated what version of old Lagos my research team could find using only these landmarks mentioned in the grants.[3] Could we locate in space what we could not locate in texts? What would it mean to interpret a land grant—in terms of direction and space—such as "below Tinubu Street" while standing on some of the fixed marks?

In July 2019, we set out to investigate some landmarks that punctuated the land grants. Some landmarks—like Faji Market, Tinubu Square, et cetera—were well known and still existed on Lagos Island. The cemetery is one of several landmarks used to orient the land grants distributed by Ọba Dosunmu in the decade between the bombardment and annexation of Lagos. In the mid-nineteenth century, it was an irregular rectangle bounded by Campbell Street, Joseph Street, Igboṣere Street, and Hamburg Street. New squares sprung up around it, and by 1885, Ajẹlẹ Square and Campos Square were depicted as forming its western boundary. Other spaces, like the fields, had been desperately fought over in the mid-nineteenth century, but had been transformed or written over by the turn of the century. These were the kinds of spaces whose traces we searched for in the urban fabric.

Macaulay's quote above points deftly to the ways life and death can coexist in the same city space. The cemeteries in Lagos Island store scripts about the past for the intrepid historian to discover, respectfully. The colonial government moved the cemetery in 1872, and acquired new land (belonging to Captain James P. L. Davies) on the eastern section of Lagos Island on what is now Ikoyi Road. It established a new cemetery there. Again, it was filled quickly, prompting the government to acquire the land opposite this site as well. We visited three sites related to what is now called the Old Burial Ground.

The conventional narrative about Lagos explains the expansion of the city in broad strokes, moving east from Isalẹ Eko and Olowogbowo to newer quarters like Ẹpẹtẹdo and Lafiaji. In this context, the land grants supply a fine-grained, if still fragmentary, sense of the pace of expansion, and indigenous sense of place in old Lagos. Yet, given this limited source base, how much can we reconstruct the way Lagos expanded eastward?

In the past, the land grant process was flexible but clear: you went to the ọba, and if you received permission for a plot, then you cleared the land, and then you built something on the land. But with the advent of documented land grants routed through the British administrators, land had to be accounted for in new ways: it had to be situated physically; it had to have boundaries; it had to be in a specific quarter. However, in the consular decade (the period between the bombardments in 1851 and the annexation in 1861)—characterized by an influx of former slaves and overlapping and competing authorities on land tenure—there was an uneasy and contested shift from communal land tenure to individual ownership. As such, proof of ownership that would survive catastrophic interruptions was required. This transition resulted in competing claims to plots of land and culminated in formal colonial attempts to settle these questions through the Land

Commissioner's Court. In this chapter, I reveal how analyzing Ọba Dosunmu's seventy-six new grants makes visible the negotiations in making space in an expanding old Lagos, according to the new and old landmarks that were used to anchor social relations in space.

GIRIPÉ

Around 1860, Giripé, a returnee from Brazil, died suddenly, leaving property behind. The land in question was a valuable commercial property, said to be "in the heart of Lagos."[4] Other records from the time that described land and property in this way were often found in the Oko Faji quarter, north of Broad Street, though the land could also have been in Brazilian Town, which was between Tokunbọ Street and Igboṣere Street. Properties in those areas, as well as in Olowogbowo and Ẹhin Igbẹti, were the subject of land grants endorsed by Ọba Dosunmu between 1853 and 1861. Regardless of the exact location of Giripé's property, his passing created an opening for the ownership of his land to be contested. Three men claimed rights to the land: Fereira, the executor of his estate; Cardosa, his former trading partner; and Agutoda, an indigenous Lagosian who claimed his father had originally cleared the land, improving it, and that thus he was descended from the original and rightful owner of the plot in question.[5]

Fereira feared that "some intrigue was being carried on" in the wake of Giripé's death. Cardosa quickly used the opportunity to secure the land for himself, likely to continue trading there. Agutoda appealed to Ọba Dosunmu that the plot in question belonged to his family, and now that Giripé was dead, the rights to it should revert to him. Cardosa claimed he did not know that Giripé had a land grant, and that they had "traded there together and paid no rent," a claim which seems disingenuous.[6] Cardosa, likely considering this local claim solid enough to establish ownership, did something unexpected: he arranged to pay Agutoda twenty-eight bags of cowries to secure his ownership of the land. This was a strange move because he must have known there were land grants in circulation.

The matter did not end there. In a stunning reversal, five years later Fereira made another claim regarding ownership of the plot; and this time, he had what he saw as incontrovertible evidence. Fereira produced for the new chief magistrate's court indisputable evidence of ownership: a land grant from Ọba Dosunmu. Either the grant had been missing since Giripé's death, or the executor could not locate the grant in 1860. This grant was ostensibly one of the estimated seventy-six land grants that Ọba Dosunmu endorsed between 1853 and 1861, when he parceled out parts of Lagos Island to strangers and locals alike. The chief magistrate found Fereira's case "fully proved" and in his favor. No official record of this grant has survived, except for the reference to it in the summary of the civil case *Ferreira v. Cardosa,* recorded in *The Anglo-African* newspaper in November 1865. Why did a land grant from Dosunmu supersede a more recent agreement with Dosunmu that Agutoda had rights to the land?

Land ownership in consular Lagos was rife with contradictions: Agutoda's initial appeal to the ọba appealed to the ọba to legitimize his claim, which suggests an investment in local traditions and understanding of the land, while at the same time, he sold the land to Fereira, effectively transferring ownership to him, or at least almost. Outside the context of the courtroom, and without British influence, the land grant would likely not have had the same power. But as instruments of British colonial and bureaucratic power, these documents conferred specific rights to the landowners, detaching them from local jurisdiction, though that was perhaps not the intention when Ọba Dosunmu gave them away.

This episode sheds light on the significance of Ọba Dosunmu's land grants during the consular period and into the early colonial period, demonstrating how land could be acquired, lost, and even transferred; how overlapping claims from native and British ideas tested land claims; and how the land grants superseded even local claims in the eyes of the British legal system in the new colony. Ultimately, the dispute demonstrates how the land grants were used and wielded as instruments of ownership, even when local land tenure was in flux.

The socio-spatial makeup of Lagos from the ascension of Dosunmu in 1853 into the early years of colonial Lagos, after its cession in 1861, defined Ọba Dosunmu's land grants. As formerly enslaved people like Giripé arrived in or returned to the city, they sought land, in some cases eventually becoming wealthy landowners, settling primarily outside the older areas in the northwestern corner of the island. While indigenous Lagosians primarily continued to occupy Isalẹ Eko, Ọfin, and Ereko—areas absent in Dosunmu's land grants—Dosunmu granted lands in new areas of settlement and commerce to the south and east.

These land grants mark a transitional period in the city's spatial politics, as the power to grant land had traditionally resided with the white cap chiefs and then, under Kosọkọ, the war chiefs. As the chiefs' role in parceling out land diminished, residents and potential residents appealed to the ọba to grant them land. But the grants required a consular seal and then, after cession in 1861, needed to be reviewed by the Land Commissioner's

Court. Competing claims around the ownership and distribution of land continued to plague residents. Embedded in the spatial history of this town is the question of how communally owned land became individual land, and how that shaped the urban fabric of the city. In answering this question, historians have turned to land tenure. In the sections that follow, I explore land in Lagos in its social and spatial form to draw new meaning from the distribution of land.

The cartographic evidence gleaned from walking the city and from consulting historical maps supports these records and points to inconsistencies and incongruences in the ways the grantees conceived of the places they wanted to control. I argue in this chapter that by providing a new language for land distribution, the grants speak to how land and land acquisition were tied to specific preexisting social relationships, while also creating the potential for future ones.

Land grants, as an example of colonial policy, have generated more interest than the spaces they bounded, or peoples who sought the titles. This chapter uses the grants as a retelling of a portion of Lagos Island. These grants make possible new visions of Olowogbowo, Oke Popo, and Oko Faji. They also draw attention to landmarks, new and old, that are spaces of competing meaning between Lagosians and the changing migrant population of formerly enslaved people, missionaries, European merchants, and British administrators. But was a new geography of the city emerging, or are the grants evidence of how that space was already being imagined and used?

For instance, we see how Olowogbowo, or the "west end," developed into a new residential area, close to the commercial strip on Water Street. Only one other street is named—Bankọle—in the grant given to Charles Wilson in November 1858. To be clear, most if not all the island was occupied, but not every part of the island was written about or made it into the archive. There was a rationale for requesting grants, and that becomes clear when they are analyzed in terms of use and then value.

Historians have explored why newcomers and some older residents sought these grants, and how private property emerged on a larger scale on the island. Given the island's history, there was no reason to assume that a new regime would not sweep away the current occupiers of any space. Just as the 1851 bombardment had displaced Kosọkọ and his retinue, another regime could potentially displace the newcomers who had arrived with Akitoye and the British in 1852. There was a precedent for this, and besides, Akitoye himself had been displaced in 1845. However, the British occupation proved durable and was able to fend off war and rumors of war around the return of Kosọkọ. One could also imagine that returnees and new arrivals wanted to secure their property.

The question of who owned the land in Lagos was a vexed one between the ọba, his idẹjọ chiefs, and, by the 1850s, British administrators. Local stories and myths suggested that the Olofin, who was the origin ruler of Lagos, divided the land and lagoon between several important men. Their names and titles indicated ownership and lent credence to this origin story—for instance, the Oloto of Oto, Onikoyi of Ikoyi, the Oniru of Iru, et cetera. These men were known colloquially as "white cap chiefs" on account of the headgear they favored.[7] Prince Eleeko (Ọba Dosunmu's son) testified in 1912 that of the four classes of chiefs, only the idẹjọ were Yoruba.[8]

A WALKING CARTOGRAPHY: MAPPING ỌBA DOSUNMU'S GRANTS

The landmarks in the grants form a prequel to placemaking in Lagos. When the streets were named, they relied on a series of landmarks already in use in Lagos by the ọmọ Eko and the ara Eko. The coming of new populations solidified some of the old landmarks, destroyed some of the older ones, and made new spaces (like the racecourse). In map 4.2, I explore the locations of this set of landmarks as they emerged in the 1850s and up to the annexation of the city in 1861. As such, the premise for this chapter is that they formed a new overlapping template from which the street network emerged in 1868. You can see these landmarks in map 4.2.

Lagos, while visible politically in the secondary sources, is spatially opaque in the consular decade. We know where people settled but have little understanding of how or why. Land grant records bring more granularity to the understanding of space in Lagos. The earliest grants given by Akitoye (and examined in the previous chapter) provide information on a handful of residents, like missionaries, merchants, or consular officials. Without Ọba Dosunmu's land grants, there are few details about the urban form in Lagos, and it is difficult to see the specific relationships between people and space outside of missionary and administrative circles.

The evidence tracing these grants is fragmentary at best. Few grants exist in their complete form. A small sample of the grants can be found in the Foreign Office correspondence written by consuls in 1855; and the text from a handful of grants is reproduced in the minutes of land-ownership investigations that were convened infrequently and the reports produced around them.[9] Martin R. Delany, an African American intellectual, visited Lagos in the 1850s, and received a plot of land from Ọba Dosunmu. He reproduced a copy of his 1859 grant to land in Oke Popo in his text *The Official*

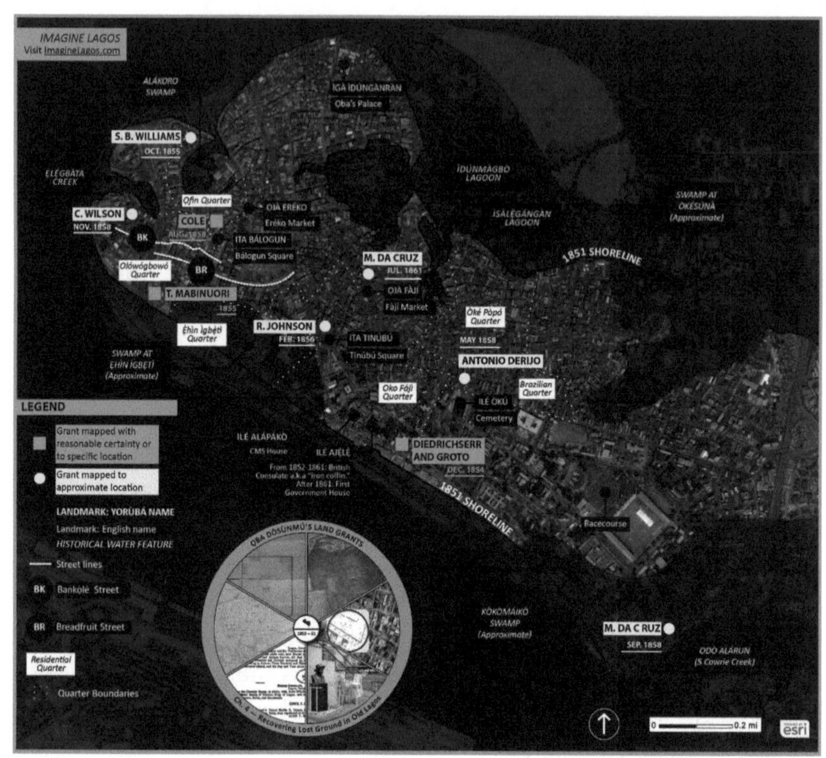

MAP 4.2. "Ọba Dosunmu's Land Grants." Base map courtesy of ESRI. Images in the cartouche are, clockwise from top right, as follows: (R1) portrait of Ọba Akitoye, in Isaac B. Thomas, *Life History of Herbert Macaulay, C.E., Third* (Lagos: Tika-Tore, 1946); (R2) Spyglass over Joseph Street and Campbell Street in Lagos Island; (R3) Campos Mini-Stadium, original site of old cemetery, Lagos Island, 2019, photograph by author; (L3) land grant to Martin Delany, 1855, from Martin Robinson Delany, *Official Report of the Niger Valley Exploring Party* (New York: Webb, Millington; London: Thomas Hamilton, 1861); (L2) detail from John H. Glover, *Sketch of Lagos River,* 1859 [corrected to 1889], 1:20 scale, MR Nigeria S.127–24, Courtesy of the Royal Geographical Society. London; (L1) plan of Church Missionary Society plot, 1855, from Slave Trade No. 6, FO 84/876, courtesy of the National Archives, Kew, UK. Map by author.

Record of the Niger Valley Exploring Party. The most complete record of the grants was created by the historian Kristin Mann, and it exists as a series of notes she took in 1985 while doing research in Lagos.[10] The grants followed three templates, so it is possible to reconstruct parts of them according to the text formulas in use. The original records in Lagos are currently inaccessible.

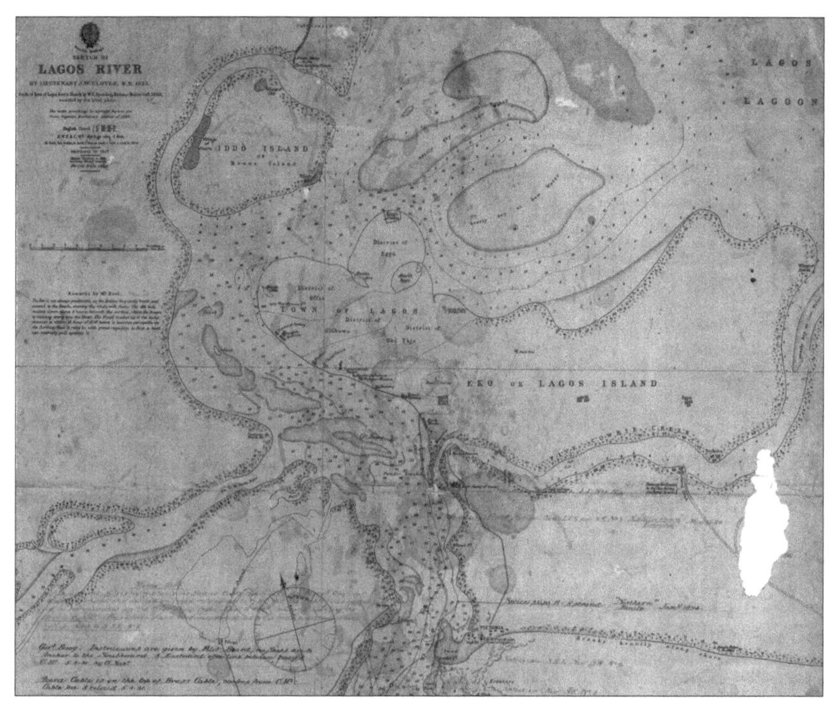

FIGURE 4.1. Detail from John H. Glover, *Sketch of Lagos River,* 1859 [corrected to 1889], 1:20 scale, MR Nigeria S.127–24. Courtesy of the Royal Geographical Society, London

Despite their fragmentary nature, these records provide the richest and most fascinating insight into the ways land was desired, understood, divided, and distributed in Lagos between 1853 and 1861. When read with other contemporary records, including newspapers, consular correspondence, claims in future court cases, and land tenure investigations, the wording of these grants provides a new grammar for the analysis of space in Lagos, as their language, diction, and intention ushered in a new vocabulary for land distribution and ownership in Lagos.

Even though some grants reference a plan of the city, that map is one of the missing maps. Here, I analyze the only known contemporary map of the whole of old Lagos: John Glover's 1859 *Sketch of Lagos River* (see figure 4.1).

Macaulay (a grandson of Rev. Samuel Ajayi Crowther) conducted the most meaningful investigation into the "land of the people" in 1912.[11] Although he is better known now as a nationalist icon, he leveraged his background as a surveyor in the Lagos colony to provide a unique interpretation of the historical relationship between land and politics in Lagos. To answer questions about the ownership of land in Lagos, he relied on four main sources: Ọba Dosunmu's land grants, the minutes of the Land Commissioner's Courts from 1863 to 1866, treaties, including the one signed at the cession of Lagos in 1861, and Glover's sketch of Lagos. Speaking of this map, he noted, "[You] will see [that] on this Chart the occupied, possessed and established portion of the Island which formed the town of Lagos is carefully demarcated, etched and named."[12]

Glover's 1859 map sets the historical and geographical context for the quarters and hints at the origins of their spatial relationship. Even though it is a sketch of the river and offers more detail about the depth of the lagoon, this map is the earliest physical representation of the districts that oral sources reference: Iga / Isalẹ Eko, Olowogbowo, Oko Faji, and Ọfin. Missing from this map are Ereko and Portuguese Town.

In map 4.2, I plot information from two types of Dosunmu grants. In one set, the locations of grants can be determined with a high degree of specificity, such as Diedrichserr and Groto's grant in Olowogbowo or Tom Mabinuori's 1855 grant for his property "at the waterside near Igbẹti."[13] These grants are indicated on the map with a blue square.

The most important data in the grants are those which are both social and spatial. We have the complete text of six grants. They include land given to (1) P. Diedrichserr and H. Grotto in Ẹhin Igbẹti in 1854; (2) James Cole "on the waterside" in 1855; (3) Thomas C. Cole "on the waterside" in 1855; (4) Le Gresley, the Banner Brothers' agent, "on the waterside" in 1856; (5) Luiza Muribeca in Oko Faji on May 5, 1858; and (6) Martin Delany in Oke Popo on October 25, 1859.[14]

This map shows all the quarters that are identifiable from the land grants that are available. From the evidence, we see that Dosunmu affixed his seal to land distributed to established districts, such as Oko Faji, Olowogbowo, and Ọfin. New types of structures, mostly commercial and mixed use, were beginning to define the waterfront, changing Ẹhin Igbẹti to the new commercial hub of the legitimate trade. Newer still were districts like Oke Popo, which from the grants first seemed to be a part of Oko Faji and then, by 1868, emerged as a site of its own. Even though there are several grants to returnees from Brazil, such as Derijo, Libeiro, and Lisboa, and there was written evidence of a thriving Brazilian quarter by 1853, there is no explicit mention of a Portuguese or Brazilian town in Lagos in the grants from the consular decade.

Other grants were given in proximity to the Christian burial ground, including a grant to Christa Alves de Souza on August 5, 1857, and to Jose

Goncalves Bartos on August 11, 1857, at the "upper part of Oke Popo at Oke Fajì below the burial ground."[15] Robert A. Johnson received land near Tinubu at Oko Faji on February 4, 1856.[16]

Mabinuori died on May 17, 1874.[17] Mabinuori's compound is described as a plot between Water Street (the Marina) and Broad Street, north of Bishop Street.[18] He was said to have acquired the property around 1851, which would mean he made a claim to it around the time of the bombardment of the city. Mabinuori is noted as being formerly enslaved in Lagos.[19] Since there was another claim that the land used to belong to Ọṣodi Tapa, it is likely that Mabinuori claimed it after the bombardment when Tapa left the city with Ọba Kosọkọ.[20]

There were striking debates on this case on the rights of women to control a family and its property, but its implications in this case are outside the purview and period of this project. In summarizing the case, Chief Justice Osborne noted that "one of the most striking features of West African native custom" was "its flexibility."[21]

The names and residences of indigenous Lagosians are rarely given in fixing the locations of these grants. Curiously, neither are those of the chiefs, whether idẹjọ or not. Akin L. Mabogunjẹ's research has shown how almost every chief in Lagos resided in or near the Isalẹ Eko quarter.[22] Since most of the grants were given outside Isalẹ Eko, it makes sense that none of them would be mentioned in fixing the locations of the new landowners.

In the second set of grants I indicated in map 4.2, the information we have is mostly social, and leaves the grants within the boundaries of a certain district, such as Derijo's grant opposite the burial ground. These grants are indicated with a yellow circle. Together, these grants show new data for the filling in of Oko Faji and the development of the Oke Popo district. They also highlight the important new landmarks like the racecourse in the east, and the new cemetery. All these new and old sites allow the grants to be indexed in space. The "field at Oko Faji" comes up in several grants. "The fields" was used as a reference to open spaces east of the originally occupied parts of the town, encompassing areas now known as Oke Popo. It could also be a reference to the racecourse, but an 1860s grant to B. Eutrado mentioned the racecourse specifically.

At least twenty-four of the recorded grants mention Oko Faji. Within this subset are grants that anchor space using landmarks like the cemetery or market and point to "the fields" and Oke Popo as part of Oko Faji. Some grants offer more information than just the district, such as

Johnson's grant dated February 4, 1856, which stated his land was "near the ground and premises occupied by Tinaboo."[23] This would put his plot near Tinubu Square.

THE POWER TO GRANT LAND

Migrants who returned from Brazil (and to some extent Cuba) congregated in Oko Faji, a growing district a few miles farther east. Grants in Ẹhin Igbẹti, on the southern edge of Lagos Island, were mixed, with Sierra Leonean returnees and European firms joining a heterogeneous community on the waterfront made up of a thin line of CMS holdings, the British Consulate, and a handful of European merchants whose establishments faced the Atlantic coast. There was also another group of grantees who were employed by the CMS, and they received grants to erect homes or mission houses. For example, James C. Barber, schoolmaster for the CMS, received land in Oko Faji "for his use" in 1858.

Most of the grants are clearly marked for men, almost all of whom were from Brazil, Europe, and Sierra Leone. There are several grants that are incomplete or missing from the records. These missing grants likely went to Black men from the Americas interested in living in Lagos. They include Joseph Harden (an African American Baptist missionary), Robert Campbell (a Jamaican), and Martin R. Delany (an African American) of the Niger Valley Exploration Party, who visited Lagos in the late 1850s. Only one indigenous Yoruba, Mabinuori, himself a long-term resident of the city, could secure grants to land in Olowogbowo. In other grants, the people mentioned were native to Lagos, such as in the case of Captain James P. L. Davies's 1859 grant, in which he received land that belonged to an Ojo (Ojoe) Oniyan.

Together, these men were part of an emerging elite group who now had "tangible" proof of possession (in the European sense of land title) that could ostensibly resist the usual challenges to land ownership in Lagos—war, slavery, and interracial and interreligious conflict. Most notably, their acquisitions were unconnected to the previous system of patrons and clients, routed through the white cap chiefs, who had until then determined the patterns of land ownership in Lagos.

But despite their focus on strangers, these grants showed how (and sometimes where) Lagosians occupied space, as their locations often depended on a recursive system that relied on networks of people and places that allowed them to mark their spaces in the town.[24] There was no easily accessible or archivable system of street names in the town for strangers, yet it is because of their proximity to Lagosians that we see glimpses of figures, some well known and some who are otherwise hard to see. These grants foreground an emerging relationship between immigrants and Lagosians.

For instance, Johnson's grant put him near Efunroye Tinubu, who was the most important woman merchant in Lagos. His grant notes he was given "land at Oko Faji," and to qualify this, it added that it was "near the ground and premises occupied by Tinaboo [*sic*]."[25] William Gomez, a shoemaker who was an immigrant from Cuba, acquired in 1861 a grant to land "at the back street of Tinubu," which likely meant north of Tinubu Square. Ironically, Tinubu herself was soon evicted from the land and square that bore her name a few months after Johnson's grant because of debts and tensions between some traders and the consul who accused her of plotting his assassination. However, this landmark and street continue to bear her name.

There are documented exceptions to the filling out of these districts, such as when Mathews da Cruz received a grant for land near Five Cowrie Creek, and Bentunato Eutrado received land near the racecourse.

Every grant in the record was given outside the Isalẹ Eko quarter. Several roads in Lagos met at Faji Market in nineteenth-century Lagos. Historically, the market was located almost due north of Ita Tinubu, and the old maps show it was located between two important lagoons: Isalẹgangan and Idunmagbo. Topographical maps show that the location was ideal, as it was on the crest of land that had the highest elevations in that region—ranging from at least thirteen to fifteen feet above sea level. In a low-lying island, even that slight elevation would make a difference.

The origin of the grant-giving process is not clear, but it followed the precedent set by Akitoye in 1852 when he granted land to the CMS. By the time Dosunmu became ọba of Lagos, the Saro had already begun to settle in Olowogbowo; Portuguese Town was already becoming a site for the Brazilian returnees; and thousands of Lagosians were crowded in the Iga / Isalẹ Eko, Ereko, and Oko Faji quarters.[26]

Ọba Dosunmu came to power in September 1853, after the death of his father, Ọba Akitoye. A month or so after he became ọba, Dosunmu gave his first grant of land in Lagos. This first grant was to James George, an immigrant from Sierra Leone. On December 12, 1853, Dosunmu gave him land in Oko Faji, where he could "erect stores and other buildings."[27] George was a Sierra Leonean merchant, and he was given land on the waterside next to James Sandeman, a British merchant.[28] Based on the data from Consul Benjamin Campbell's letters, this plot was most likely at the intersection of Apongbon Street and Water Street.

This single document represented the end of a negotiation that must have mirrored the older patterns of acquiring land on the island, as evidenced by the struggle with the CMS in 1852.[29]

But why was Dosunmu suddenly giving land away in this way, in contrast to older processes that involved the permission of the land-owning chiefs? Lagos's ọbas had been under pressure since 1852 to provide new settlers with formal documents that guaranteed ownership of desirable plots of land. For instance, in 1853, during the dispute between Consul Campbell and the agents of the CMS, the Foreign Office had instructed Campbell to secure a grant that proved the ownership of the land the British Consulate was built on.[30] It is also possible that Dosunmu, like his father before him, was under the sway of the CMS agents, and this process of fixed grants was a way to ward off the challenges the Christians had faced when they first moved to Lagos. What is clear is that the process depended on Dosunmu's perceived jurisdiction over the distribution of land, a claim in sharp contrast to the historical or "native tenure" of land distribution that was channeled through the chiefs.

After the bombardments, land had been guaranteed by treaties agreed to between the merchants and the king. These new grants by Dosunmu were imagined and positioned to maintain ownership and solidify the first large-scale partitioning of land to various individuals who intended to make their homes and fortunes on the island.

Each grant was dated and signed in Lagos at the British Consulate, with the British consul as a witness. Dosunmu hired Turner, a Saro man, to measure the plots of land he granted.[31] The grantees' full names and place of origin were indicated, and their proposed use of the land was registered. Often, there was also a brief description of the plot, with its dimensions and location indicated as precisely as possible or necessary. Most of the grants have not survived in their original form. However, in 1897, John J. C. Healy (then the commissioner of lands) carried out a thorough investigation of the process of grant giving in nineteenth-century Lagos.[32]

Three accounts of the process of acquiring grants have survived. As accounts of the process of strangers gaining land in West Africa, each retains a unique framing of the event. Delany, an African American visiting West Africa, provides the only existing contemporary record, in which he described his own experience of acquiring land from Dosunmu in 1859. His account is framed around the opportunities possible for others like himself who would be interested in moving to West Africa. The other two versions come from the late nineteenth century and early twentieth century from

Macaulay, a former surveyor and famous nationalist, and Healy, a land commissioner. Macaulay framed his account within a report on "the land question" in Lagos, especially as it pertained to how much of the land the British administration had acquired via the cession in 1861.[33] Healy's account was in the report on land tenure, part of the colonial administration's attempts to adjudicate land cases while maintaining their own hold on the territory of Lagos and beyond.

Delany arrived off the coast of Lagos in September 1859 and was escorted to the city by Davies, a prominent Saro merchant. He spent five weeks in the city, during which he made the acquaintance of several chiefs and important persons; he was "waited on by the messenger of the king." He also noted that he had visited every important place. After some deliberations, he was granted land in Oke Popo (for him and his heirs) in a document that was "duly executed, signed, sealed and delivered," but only after the land in question had been measured in his presence, with Joseph Harden as a witness.[34]

Healy's version of the events was like Macaulay's. Parties interested in acquiring land had to approach Ọba Dosunmu, who would then give his staff (a symbol of his approval) to a messenger, who would supervise the finding, measuring, and giving of the land. Though land was technically never sold, there was still a financial aspect to the transaction. Each applicant was required to pay ten heads of cowries (the equivalent of two pounds and six shillings) for the messenger's services. This fee was standard, regardless of the size or dimensions of the plot. After this, the applicant would receive the grant document from the clerk engaged by the British consul, and this document had to be taken to the consulate for "ratification" via a seal affixed by the consul, implying that the document was not fixed or fully binding without the input of the consul. Finally, after a few days, the clerk would affix Dosunmu's circular stamp on the document as proof of its authenticity, like official items such as deeds, letters, and documents. This "deed of land" was noted as registered in the Registry Book in the consulate.

Macaulay's interpretation of the process involved the ọba, chiefs, and a surveyor. In his version, if a Sierra Leonean such as Thomas C. Cole was interested in land in 1855, say, along the waterfront, he would first approach the king, who would inquire which district he desired. After naming the spot (in his case recorded as "near the waterside called Ebetee"), Ọba Dosunmu would send for the "chief of the district," who would then allow for land to be "allocated" in the area.

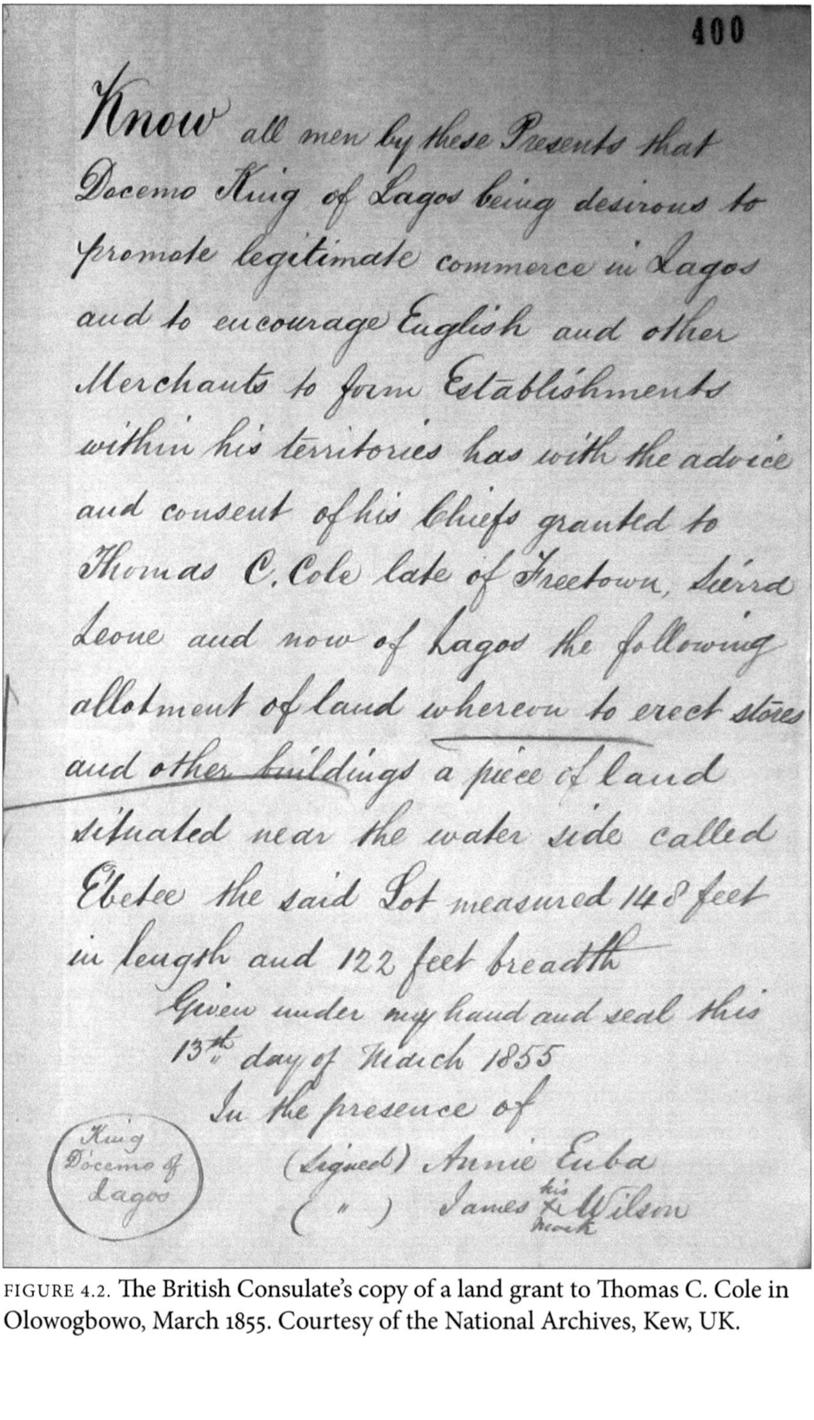

400

Know all men by these Presents that Docemo King of Lagos being desirous to promote legitimate commerce in Lagos and to encourage English and other Merchants to form Establishments within his territories has with the advice and consent of his Chiefs granted to Thomas C. Cole late of Freetown, Sierra Leone and now of Lagos the following allotment of land whereon to erect stores and other buildings a piece of land situated near the water side called Ebetee the said Lot measured 148 feet in length and 122 feet breadth

Given under my hand and seal this 13th day of March 1855

In the presence of

King Docemo of Lagos

(Signed) Annie Euba

(") James his X mark Wilson

FIGURE 4.2. The British Consulate's copy of a land grant to Thomas C. Cole in Olowogbowo, March 1855. Courtesy of the National Archives, Kew, UK.

Cole would then be escorted by a messenger who would hold the king's staff, and the applicant would come into possession of the land with both the chief's and king's consent. Cole's grant was approved on March 13, 1855 (see figure 4.2). He was given a lot measuring 148 feet in length and 122 feet wide to "erect stores and other buildings."[35] His witnesses were Annie Euba and James Wilson, who signed and made their mark. The only payment was for the "services of the staff" and a fixed "stamp fee" that went to the consul. At the successful conclusion of this process, Cole would have brought gin, rum, silks, tobacco, et cetera, as presents to the ọba, and the ọba would share them with the chief.[36]

It is not clear how consistent this process was, but an examination of the content of grants reveals that the second through the eighth grant bore the seal of the consulate. One thing is clear though—the white cap chiefs were never recorded as being involved in this process, even though they were frequently acknowledged as the "rightful owners of land in Lagos."[37] There is no evidence of the number of applications, and how many were approved or denied, and what the criteria for such decisions might have been.

"UPPER OKE POPO AT OKO FARJI": A NEW VOCABULARY FOR AN INCREASINGLY FIXED CITY

Healy's report briefly analyzed the twelve grant formats in use in Lagos from 1853 to 1897, a period that saw land in Lagos distributed by Dosunmu's grants (1853–61), Crown grants (1863 onwards), Glover Tickets (1868–78), and, finally, Permissive Occupancy Grants (1881–88).[38] Dosunmu's grants exist in the first three forms, which Healy classified as "A," "B," and "C." Of the seventy-six grants, one was written in the "A" format, nine were in the "B" format, and sixty-three were in the "C" format.[39]

The written format of Dosunmu's grants changed over time, suggesting certain patterns in the desire for the acquisition of land. The first grant to James George in December 1853 is the format Healy marked as "C," and this eventually became the most popular format for the grants in Lagos, with approximately sixty-three grants following this same template. Groto and Diedrichserr's grant, which was signed a year later in 1854, is the only one in the "A" format, and it is distinct from the others because it is the only grant given to a European and specifies that the grant was given as a free donation.

There are nine grants that appear in the "B" format, and they are distinguished by their emphasis on the promotion of "legitimate commerce."[40]

Unsurprisingly, land grants in the "B" and "A" formats are concentrated in Ẹhin Igbẹti, as most of the commercial houses in Lagos were set up to take advantage of this area's proximity to the sea as well as its distance from the "native" quarters that were found in the northwestern parts of the island in districts like Iga / Isalẹ Eko and Ereko. The next six grants are in the "B" format, with land distributed in Ẹhin Igbẹti (five plots) and one in Olowogbowo. In this format, we see the following template:

> Whereas Docemo, King of Lagos being desirous to promote legitimate commerce in Lagos and to encourage English and other Merchants to form establishments within his territories has with the advice of this Chiefs granted James Cole of Freetown and now at Lagos the following allotment of land whereon to erect stores and other buildings a piece of land situated near the water site called O'Keate's [*sic*] wharf the said lot measured 108 feet in length and 1414 feet in breadth.[41]

Then, in October 1855, the rhetoric returned to the "C" format when Walker Strober and William Marsh received land "near Ọffin" and in Olowogbowo, respectively, in order to commence their commercial activities.

With these grants, it is possible to draw a portrait of four significant developed areas in Lagos, namely Olowogbowo, Ẹhin Igbẹti, Oko Faji, and Oke Popo, as well as landmarks that came to hold increasing significance in the city, namely the burial ground and the racecourse. Important landmarks had also developed around the perimeter of the island in the areas that fronted the Lagos Lagoon (often marked as the "waterside"), Balogun Square, and Tinubu Square. The older sections of the city were absent from the grants, and are the areas where the older, indigenous population settled.

In these grants, there are several important landmarks: the racecourse, the lagoon (or waterside), and Five Cowrie Creek (the small body of water that divides Lagos Island from Kuramo Island). The grants use specific markers in Lagos, relying on places, people, and features that have acquired a certain permanence of place and, therefore, meaning and are subsequently used in plotting the locations of these grants. So, for instance, when in James George's grant, we see his plot as situated "next to J. G. Sandeman," or in Bernado's case, "near that of Tibeiro Emmanuelle Gomez in the field of Oko Faji, opposite the burial ground," we begin to see how these series of references plot a tangled web of spatial relationships in Lagos.

Another important landmark that features prominently in the Dosunmu grants is the burial ground. There are few references to it outside of the grants, but, in fact, its existence was planned as early as 1852, when Article VIII of the agreement with Ọba Akitoye declared the following: "The King and Chiefs of Lagos further agree to set apart a piece of land, within a convenient distance of the principal towns, to be used as a burial-ground for Christian persons."[42] Even though this burial ground is an important landmark in several of the grants, it cannot be located with any certainty until the 1883 sketch of the town and island of Lagos, where it appears in visual form for the first time.[43] Another important feature is its planned distance from the town. In this grant, we see the importance of establishing places outside of the "native" dwellings, but by the end of 1861, we see that Eko and Lagos have become a single entity.

The specifications of several grants are unclear and contradictory when read now, but they undoubtably conveyed specific and verifiable meanings when they were written. [44]

The problem of transliteration and translation poses interesting questions for interpreting space in midcentury Lagos. Even though the Bible had been translated into Yoruba since the 1850s, and Reverend Crowther had also published editions of his Yoruba-to-English dictionary since the 1840s, there was no written consensus on how to render place-names on the island.[45] So, on the maps and in the correspondence about Lagos, there was a marked variation in how space was recorded, until 1883, when place-names took on the current uniformity. There were no standardized systems of names with which foreigners to Lagos mapped space in the city. Even the name of the island was in question. For example, until 1883, we still see Eko on the maps of the city, though it disappears by 1885. Official British correspondence rarely referred to the island as Eko, unless speaking of the indigenous quarters.

The Oke Popo area of Lagos is fairly neglected in the historiography of mid-nineteenth-century Lagos, but it features prominently in the grants given by Ọba Dosunmu. One of the major problems with understanding the specific locations of the grants is the lack of a consistent system of street addresses in Lagos. Compounding this is the fact that there were no agreed-upon transliterations for the names of the different indigenous districts, and each correspondent made an approximation of the names of the areas referred to in letters, grants, and treaties. Thus, Olowogbowo became "Olubowo" on maps, and "Olowobo" or "Oloboo" in some letters and grants; Oko Faji was alternatively "Farji," "Fajee," or "Faje"; "Okai Po Po" was definitely an attempt to spell Oke Popo; and so on. These were the easier ones to render. The grants are filled with confusing references to plots, such as "at Oko Fájì called Ebetu" and at "Upper Oke Popo at Oke Faji below Burial Ground." Based on an examination of Glover's 1859 *Sketch of Lagos River,* a map contemporary to the period of these grants, as well as John Pagan's and William T. G. Lawson's sketches of Lagos in 1883 and 1885, respectively, the former likely refers to land in Ẹhin Igbẹti, while the latter most likely refers to land immediately south of the cemetery, putting Christa Alves de Souza's grant in the vicinity of the eastern extreme of Ẹhin Igbẹti. However, despite the difficulties in translation, the grant introduces and places an interesting set of landmarks in the city.

The allotments of some locations seem to be entirely unclear—for instance, in the case of Jose G. Bartos, who received "ground in the upper part of Oka Popo at Oke Faji below the burial ground." His document references three overlapping yet distinct reference points on the island, if read from a perspective that relies on a north–south orientation for plotting points that are above or below another. Later maps and other grants indicate that Oke Popo is an area distinct from Oke Faji, and maps of Lagos locate the burial ground.[46] On first glance, the area specified seems to refer to two different districts, but subsequent grants, such as to Joao Aucelus, point to the same area: "ground in the upper part of Oke Popo at Oko Farji."[47]

Almost a year after the first grant, on December 8, 1854, Dosunmu gave a plot in the Ẹhin Igbẹti to two merchants from Hamburg: "Docemo, [*sic*] King of Lagos, certifies that he has given Messrs. P. Diedrichserr and H. Groto the piece of land adjoining the Wesleyan Yard to the north-westward measuring 8000 square yards."[48]

These grants are at once illustrative and iterative in their formation. They are illustrative in the sense that they show that Groto and Diedrichserr paid nothing for their land, and that we know the overall area that they received. However, they are iterative because of the circumstances of the spatial logistics of the island. Without a system of consistently "named" streets and "exact" or Cartesian locations amenable to the granter and grantee, no grant could stand on its own, and no piece of land could yet be located without a reference to a previously existing landmark or location. Thus began a project of fixing space based on iterative grants that built a spatial system based on a growing web of landmarks, streets, quarters, and locations that would eventually form the basis of the new shape of Lagos. The first of these grants are vague and depend on the specific points of interest to gain a foothold on the island, but by the end of the process, a more formal language of location emerges to fix land ownership in Lagos.

The readings below are according to the existing grants. Even though there are only eight land grants for Olowogbowo (dated between 1855 and 1860), it is clear that by 1855, Olowogbowo was the most popular area for Sierra Leonean returnees.[49] In a letter to the Foreign Office, the consul declared that that quarter was made up entirely of Sierra Leonean families.[50] The question remains as to how they acquired this land, and the only evidence is in a letter where Gollmer claims he simply gave them the land.

So much of what is promised in the 1854 grant to Groto and Diedrichserr clearly favors their own interests over Dosunmu's. After this grant, his pace of endorsing grants quickened, and by the end of 1855, he had affixed his seal to twelve grants, then eight in 1856 and another twelve in 1857. The year 1858 has the distinction of having the most recorded land grants, as Dosunmu gave more than double of the amount from any year previous or after it. He distributed at least twenty-five grants from Olowogbowo to Oke Popo, leaving only Ọfin untouched that year. Fourteen grants that year were given in the Oke Popo area. The pace of grant giving reduced in 1859, as that year there were only five grants, given in Olowogbowo and Oko. In the last eighteen months before the cession of Lagos on August 6, 1861, Dosunmu only gave out eight grants.[51]

The two most interesting sections of grants are in Oke Popo and Oko Faji. It is this boundary that makes the granting of land in Oke Popo so striking, and the evidence suggests that these grants were taken outside the city. Oke Popo comes into clear focus for the first time because of the nine grants given between 1857 and 1859.

How did Eko's population expand in the nineteenth century? One historian of the Oke Popo district, Alhaji Adams, dates the history of settlement to Lagos in six waves. The first was in the Isalẹ Eko, Idunmagbo, Itọlọ, and Ebute Ero quarters. The next was from the south of Idunmọta to the Marina. The third wave moved to Lafiaji, and the fourth to Aroloya, Anikantamo, Oko Faji, and Massey Square. Next was east of Tinubu into Igboṣere and Bamgboṣe up to Cow Lane between 1854 and 1859, and, finally, the Epe returnees who came back to the city with Kosọkọ and Tapa in 1862.[52] Although interesting, this compressed narrative leaves out important districts like Olowogbowo and Ereko.

According to one origin story, the name Oke Popo is derived from the first settler, Baba Okepopo, who was a woodcutter. He built his home in the hilly area (hence the *oke* in the name) said to be between present-day Ọdunfa Street and Ajanaku Street. Another version maintains the name comes from a large tree that was eventually called the Popo tree and under which important meetings were held. It, too, appended the hill to the name. Hence, Oke Popo.[53] Relative to the swamps and lagoon adjacent to this area, Oke Popo is a few feet higher in elevation.[54]

THE LAND COMMISSIONER'S COURT

The British government annexed Lagos Island in August 1861. I explore the implications of this seizure in more depth in the next chapter. However, this action had a specific outcome on the distribution of land in Lagos. After the cession in 1861, these grants were "recalled" and a new system was introduced.[55] The cession prompted the arrival of a new governor, Stanhope Freeman, who was given broad latitude with disposing of land in the city. He issued a select number of grants, often denying grants to Sierra Leonean immigrants, who did not actually reside in the city yet, in favor of those who were seen to be actually "improving" the land that they had already occupied or had been granted.[56] Of course, some of those who were denied land lodged complaints with the Colonial Office. After the cession in 1861, questions of land ownership in Lagos came to the fore. Again, who owned the island? Did the British own the entire island, and did Dosunmu even have the right to cede it to them?

In reporting on the use and distribution of Lagos since the advent of the Dosunmu land grants, Henry Stanhope Freeman was sure that the usual process of distribution and acquisition had gone awry. Concerning "bush lands," he was sure the "native Governments" had always distinguished between land that was privately held and land that was for individual use. In fact, he wrote, "one could not claim land as his private property until he has cleared away the bush from it and turned it to some account."[57]

Freeman reported that hiring Turner to measure out the land was evidence of a mistake on his part. Freeman suspected Turner's motivations and accused him of "underhanded dealing." The result, in his eyes, was that there were now several absentee owners in Lagos, as several Sierra Leoneans who had never set foot in Lagos now claimed to own land based on the Dosunmu grants.

Freeman was of the opinion that these grants were practically worthless, especially because Dosunmu had been known to take away land from those who he thought had not used it fully and give it to a different applicant. This added an extra layer of confusion to a process complicated by the fact that Dosunmu could not read and only fixed his authority to sale and certificate with the aid of an "official stamp" bearing his name. In addressing this issue, Freeman believed, however, that any land that had been occupied should remain with the current owner, as "native custom" would recognize his or her occupation as giving "the settler a right of property."[58]

A ninth ordinance was enacted in the colony in 1863 to "ascertain the true and rightful owners of land within the Settlement of Lagos," which led to the conversion of Dosunmu's grants to Crown grants.[59] This panel appointed three commissioners, Henry Eales (colonial surgeon), Thomas Mayne (chief magistrate), and William McCoskry (acting governor).[60] They were given access to the information on the seventy-six grants and had the power to summon witnesses or access other records as needed. The evidence suggests that they "favored claims based on grants from Dosunmu, demise by will, purchase, gift, and long occupation, sometimes unsupported by any written documentation."[61]

Every day, for around four weeks, there was a two-hour window for successful grantees to pick up their grants at the colonial secretary's office on Water Street (see figure 4.3).[62] The results of successful petitioners for the Land Commissioner's Court were often published in *The Anglo-African* newspaper in order to let people know their grants were ready and to also "enable persons, if so inclined, to dispute the claims."[63] Even though the complete notes of the commissioners did not survive in the archival record, the first stage of their work can be reproduced through their public announcements in the conversion of the Dosunmu grants to Crown grants.

These conversions are preserved in *The Anglo-African*, where, beginning in late October 1863, William I. J. Chapman, clerk of the Land Commissioner's Court, announced that the grants had been examined, investigated, converted, and were thus ready for collection beginning Monday, November 2, 1863.[64] In that notice, forty-five recommend grants were announced, providing for a cross section of men in Lagos society. Two chiefs—Kakawa and Aṣọgbọn—received grants to land along Water Street, and Henry Eales, a commissioner, received land in Oko Faji.[65]

We have a partial list of successful conversions because of Macaulay's investigations in 1912. He noted that the governor issued Crown grants instead of certificates to successful grantees. He also added that there was "no claim" made in any of these investigations as to the ownership of Lagos Island, despite the cession. In a way, he argued, the Crown grants could be seen as "protective titles" that were issued to recognize the land rights that were "previously acquired by the Grantee."[66] In the minutes of the Land Commissioner's Court, there is evidence of some of the criteria outlined above, and a map that had all the plots in question labeled. In 1864, G. Johnson replaced Eales on the committee.

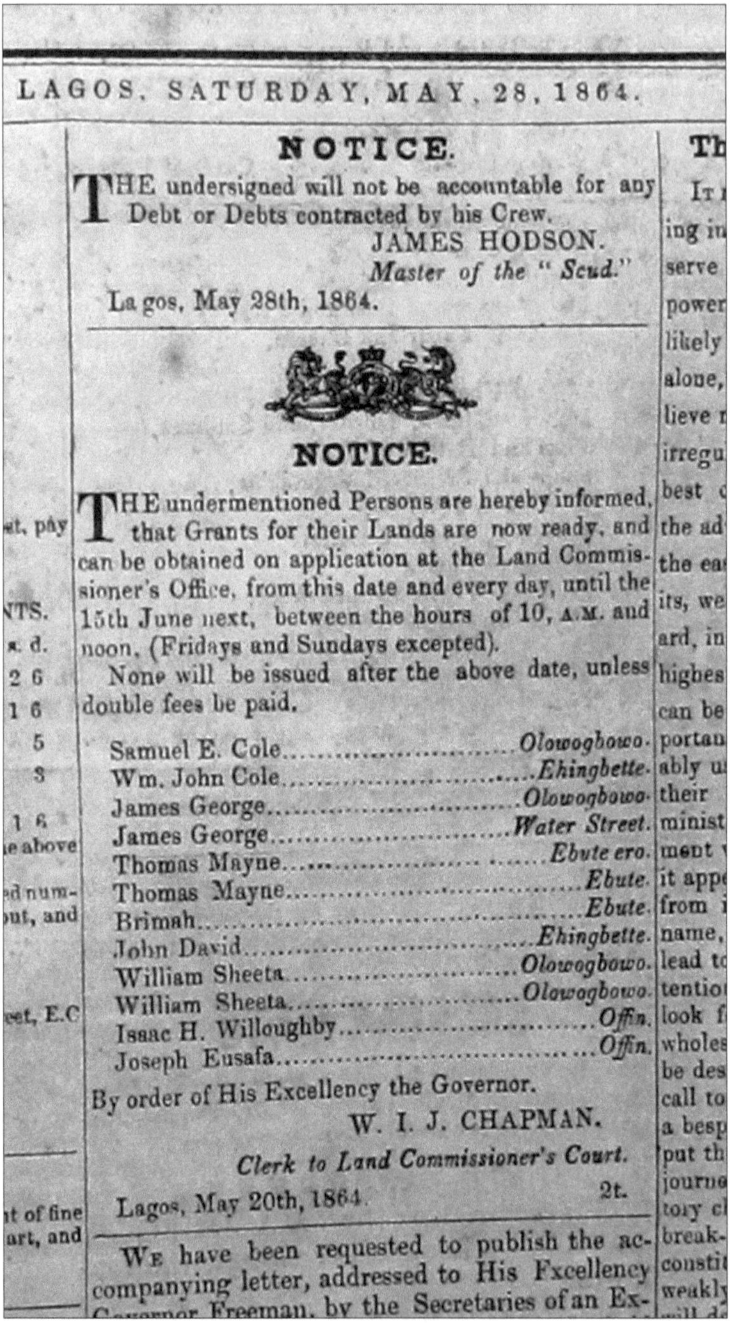

LAGOS, SATURDAY, MAY, 28, 1864.

NOTICE.

THE undersigned will not be accountable for any Debt or Debts contracted by his Crew.

JAMES HODSON.
Master of the "Scud."

Lagos, May 28th, 1864.

NOTICE.

THE undermentioned Persons are hereby informed, that Grants for their Lands are now ready, and can be obtained on application at the Land Commissioner's Office, from this date and every day, until the 15th June next, between the hours of 10, A.M. and noon, (Fridays and Sundays excepted).

None will be issued after the above date, unless double fees be paid.

Samuel E. Cole.............................*Olowogbowo.*
Wm. John Cole.............................*Ehingbette.*
James George.............................*Olowogbowo.*
James George.............................*Water Street.*
Thomas Mayne.............................*Ebute ero.*
Thomas Mayne.............................*Ebute.*
Brimah.............................*Ebute.*
John David.............................*Ehingbette.*
William Sheeta.............................*Olowogbowo.*
William Sheeta.............................*Olowogbowo.*
Isaac H. Willoughby.............................*Offin.*
Joseph Eusafa.............................*Offin.*

By order of His Excellency the Governor.

W. I. J. CHAPMAN.
Clerk to Land Commissioner's Court.

Lagos, May 20th, 1864. 2t.

We have been requested to publish the accompanying letter, addressed to His Excellency Governor Freeman, by the Secretaries of an Ex-

FIGURE 4.3. Announcement of grants in *The Anglo-African* newspaper, 1864. Courtesy of the National Archives of the United Kingdom.

When Thomas B. King claimed he had occupied the same plot in Olowogbowo since 1852 and built a home and fenced it, his grant was recommended. The Savages—William, Ephraim, Benjamin, and Joseph—also received a recommendation, as they claimed their four lots, likely in Olowogbowo and Oko Faji, which they had also built upon. Richard Macaulay claimed via long occupation that he had occupied a plot in Olowogbowo and fenced it with bamboo, but had never applied for or received a grant for the property. One Aboke claimed he had occupied his own plot in Oko Faji since the civil war in 1853. John Thomas made the same claim on Balogun Street, using the war as a time stamp. In 1866, Charles Turton replaced Frank Simpson, recommending grants for Robert Coker in Oko Faji, who claimed he received the land in 1844 as a present from Chief Faji and Abdulmain, who had owned land in Idumọta since 1860 that he received from Chief Onisemo.[67]

~

After 1861, the land grants and conveyances in Lagos multiplied rapidly, but ironically, the records are less useful, as they have not been accessible to researchers for years. And especially after the recent burning of the High Court in Igboṣere, it is highly unlikely that more information will be available to researchers soon.

The last grant on record differs from those that precede it. Francisco Magbọla (another immigrant from Brazil) received land in Oke Popo on December 24, 1861, but instead of being signed by Ọba Dosunmu, it was signed by McCoskry, acting governor of Lagos. When Ọba Dosunmu protested in 1863 that "Mi o fi ilu mi fi torreh" (I did not give my land away), he was making a case against the British annexation of Lagos in August 1861.[68] Even though he and his chiefs maintained that according to native laws, they could not sell land, by the time of the cession, they had already given away most of what was to become the most valuable regions of the island and, in almost all cases, for free. These grants are a small but important snapshot of how people occupied Lagos in the 1850s and early 1860s. They tell us how strangers gained access to land to build their homes and stores, and how they understood how space was used and valued in the city.

5 ~ Placing Justice

At Broad and Prison

> The great cry in Lagos now is I'll shoot you and pay 10 pounds.
>
> —Letter to the editor of the *African Times*

I FIRST WENT to Broad Street and Tinubu Square with questions about place making and gender. As discussed in chapter 1, Broad Street can be described by its name: it was a wide thoroughfare that eventually reached across the length of old Lagos. In the mid-nineteenth century it stretched

MAP 5.1. Spyglass over Prison Street and Broad Street in Lagos Island. Base map and Spyglass courtesy of ESRI. The historical map is a detail from *Plan of the Town of Lagos—West Africa (in 15 Sheets)*, 1891, 1 inch to 88 feet (scale), 63 × 90 cm, CO 700/Lagos 14, courtesy of the National Archives, Kew, UK. Map by author.

from Olowogbowo in the east, through Tinubu Square in the heart of Lagos, and ended just south of the racecourse. Its original name, Back Street, also cemented its orientation to the indigenous city: it was the back of the new colonial edge of Lagos.

In many ways, Tinubu Square (or Ita Tinubu in Yoruba) was the heart of old Lagos. There is no evidence of when it was first settled, but by the 1860s, it functioned as a spatial, political, and cultural hub, where at least nine streets met.[1] You could get to almost any quarter in the city from there. In old Lagos, the Ẹhin Igbẹti quarter formed a triangle-like shape with Water Street as the longest side and Tinubu Square as one of its vertices.[2] Tinubu Square divided Broad Street into the remaining two sides of this triangle. (See map 5.2.) The western side of Broad Street still forms the southern boundary of the Ereko quarter, and continues past St. Paul's Church into Olowogbowo, where it meets Apọngbọn Street. The eastern half of Broad Street stretched into the Faji quarter, and then extended in the open area, often called "the Fields" in the 1860s. But in making this eastern route, Broad Street connected the courts at Tinubu Square—police court, criminal court, petty debt court, and slave court—to the debtors' prison in Kokomaiko.[3] Broad Street looms large here, as it was originally constructed to be a wide buffer between the colonial edge of the island and the rest of the local population.

Several important streets branch off this section of Broad Street. Oil Mill Street, for example, connects to the Broad Street axis. Robert Campbell's *Anglo-African* newspaper was published there, likely during its entire run from 1863 to 1865.[4] The site of the debtors' prison was on Broad Street, at the current site of the old Government House building. It was also close to, or identical to, the site of the former Kokomaiko, and is only a few minutes' walk from Tinubu Square. If representation of Tinubu's debt was enough grounds to exile such an influential trader from Lagos—even though there was ostensibly a debtors' prison—how else were ordinary and less-well-placed Lagosians treated on the island? Our research yielded few spatial clues to think about gender and space, but a turn north on Prison Street toward Freedom Park provided more information about space, place, and incarceration on Lagos Island.

This chapter explores the spatial, cultural, and criminal significance of the transformation of Back Street into Broad Street in the Ẹhin Igbẹti area in the first dozen years of British colonial rule, and the reframing of the Ẹhin Igbẹti area into the Marina. It offers a retelling of space in the southern area of Lagos Island, as defined by the Broad Street axis, which was designed to differentiate the European and African sections of the city. The retelling of city space happens through the newspapers and maps of the time. Who could occupy which spaces freely, who was limited to which spaces, and when could people be confined? And what is the significance of the divide between public reports and private administrative correspondence and of how this divide played out in the streets of Lagos?

While the earlier walking research in this area in 2018 and 2019 was about trying to link gender to missing features, religion, the bombardment, spatial segregation (based on sanitary records), and the expansion of residential districts (based on interpreting land grants), by 2021, I was interested in questions about justice, incarceration, and the history of policing in Lagos.

The history of incarceration in Lagos often begins at the old colonial prison and centers on Freedom Park: one of the newest historical retellings of space in Lagos Island. The park was commissioned in 2010 as part of the fiftieth-anniversary celebrations of Nigeria's independence. It was built directly on top of the site of the colonial prison and was designed as a sort of late resistance to colonialism. A recent publication from the park says, "This is a symbolic re-enactment of that journey, as the old Broad Street Prison—a colonial fortress of control—is . . . dismantled to usher in the dawn of Freedom Park."[5]

But the history of colonial incarceration in Lagos dates to a time before this prison, and is linked to slavery, race, the rise of the "legitimate" trade, and the rise of debt bondage. These ideas of justice, injustice, and confinement in Lagos were best illustrated along Broad Street, in the walk between Tinubu Square and the site of the old debtors' prison at Kokomaiko where Broad Street ends, south of the former racecourse.

In the 1860s, however, Broad Street, from Tinubu Square to Prison Street, spatially symbolized colonial justice, or rather, for many Black Lagosians, colonial injustice and maladministration in the colony.

It is no accident that these sites were linked to judicial action in the new colony; physically, Tinubu Square and the courts were linked directly to the prison through Broad Street. Between them was Government House, where administrators William McCoskry, Henry Stanhope Freeman, and John Glover held sway over Lagos from the cession in 1861 to 1872, when Glover was removed as governor.

There is an implicit sense in the public memories of old Lagos that the mid-nineteenth-century city is one that is Glover's creation. In the same way that Lord Lugard is remembered as the "creator" of Nigeria (since he led the amalgamation efforts in 1914), Glover (a naval officer who functioned as both administrator and cartographer) occupies a hagiographic role in nostalgic Lagosian history. The latter part of old Lagos's history, from 1861 to 1872, is the time of his most dramatic influence on the island and the interior. This influence also played out in his continued relationship with ex-ọba Kosọkọ and Ọṣodi Tapa until their deaths (in 1872 and 1868, respectively) and his departure in June 1872.

"British Lagos" was supposed to be a better place now, according to the new administrators and their local supporters. After the 1851 bombardment, Lagos was no longer a "slave town" and, after the 1861 annexation, no longer just an African city. But what was the real experience for Lagosians now? In what ways was it better? And in what ways did the Lagosians say it was worse? Many of those answers hinge on race, culture, and place. The cartographic evidence from the period is often racialized, including maps that are careful to distinguish between the White and Black or European and African population. These distinctions are also mirrored in the only sources that collect and reflect local views: the newspapers that circulated on the west coast of Africa.

Infrastructural improvements to the city in the name of civilization and legitimate colonial commerce masked the violent racial logic of colonialism: African men and women in Lagos were subject to terrible and unequal treatment in the streets and under the law. Children, especially young girls, were not exempt from this treatment, and the cases examined below demonstrate colonial contempt for African lives. This chapter explores the logic and consequences of remaking Lagos. Were the new streets, lamps, and other issues really a panacea for the violence, violations, and exploitation of people and place?

Debates about space, race, and belonging inflected issues of justice in colonial Lagos between 1861 and 1872. Spatially, public debate about the British response to an attempted insurrection by former ọba Dosunmu in 1863 (and the elision of educated and Christian Africans from that response) and several high-profile court cases illustrating racial contempt for Black Lagosians highlighted the sharp line along the Broad Street axis that effectively separated the Marina area and its largely British inhabitants from the rest of the settlement on the island.

Although the city's colonial society was complex with regards to language, education, religion, and place of origin (as illustrated by the fallout from the 1863 Dosunmu rebellion), racial division and prejudice—symbolized by the Broad Street axis along which the institutions of colonial justice were located—were evident to African observers of the cases of James Cole,

Thomas Lubley, William Turner, and Wilhelm Rascher, as well as that of the steamer *Thomas Bazley*. Even though colonial administrators promoted colonial Lagos as a bastion of freedom, civilization, and legitimate trade, this seemed not to extend to the application of equal and just treatment of Black Lagosians.

This chapter considers these questions through four episodes in colonial Lagos, all of which hinged on place, justice, and the conditions for civilization in the new colony. "Africa for the Africans" was one claim, but who was this new Lagos built for? I show how "improving" the space produced only a patina of "civilization," and how, in fact, this new city on the edge of Lagos Island was not imagined for Lagosians, or even other Black people, African or otherwise, unless they fit in specifically with the needs of the British administrators. Even when they did, their race predicated their treatment by Europeans. These spatial connections between the newspapers, prisons, squares, and former courts along the Broad Street axis in old Lagos prompted a return to the colonial documents to reexamine their connections. As I did so, more protagonists—especially women and young men—emerged as the victims of differential justice carried out in Lagos.

In the sections that follow, I first map the data that are available on the Broad Street axis that came to divide the city. As part of the walking cartography, this map acts as a source, as a methodology, and, finally, as an illustration of the chapter's argument. Following this section is an analysis of Ọba Dosunmu's rebellion against the British and futile attempt to reclaim the city in 1863. Events from 1865 frame local attempts to control space and analyze administrative attempts to quash them. It ends with the 1869 wreck of the steamer *Thomas Bazley*, the consequences of the abuse of Africans in the colony, and the removal of Glover as governor of the colony in 1872.

In their colonial reports, administrators boasted that Lagos was now more than just an economic hub. In fact, the predictions of the postbombardment era were bearing fruit: within a decade of formal colonial rule, Lagos was a "centre from which civilization has radiated for many miles around."[6] But publicly, the newspapers, the letters, and editorials pushed back against this so-called progress, especially because of the toll it was taking on the Lagos population and on those in the Yoruba heartland.

British administrators and governors—from McCoskry to Freeman to Glover—spent the first dozen years of formal colonial rule trying to reform Lagos but were hampered by tight budgets and situations arising from surrounding territories. Dosunmu, the ọba, or ex-king as he was styled after annexation, was mostly a spectator in this process, at least for major decisions about infrastructure, justice, and the economic aspects of instituting "legitimate commerce." If Lagos had spent the 1840s being a "slave town" and the 1850s trying to shed this designation, then the 1860s was the age of a new city, remade in the image of European "civilization." Within a few years, Freeman was replaced by Glover. As the administrator, Glover enacted ordinance 17 in August 1864 to "provide for the laying out of the Town of Lagos in Broad Streets, Roads, and Highways in the Settlement of Lagos."[7] However, to construct these roads, he needed to destroy several preexisting houses. "Native" agency in preserving their "crooked, narrow streets" and "filthy hovels" was often framed as being anticivilization, and the missionaries and Europeans were often convinced that such agency lacked the understanding or preference for improvement.

The idea of establishing a "legitimate trade" had a spatial and tangible impact on the redesign of Lagos; as one letter writer said, "[Glover] has made Lagos habitable, and now he is lighting, as well as striving to enlighten her. Streetlamps in Lagos! Look at that, ye old slave-traders, who may still be brooding over past atrocities of your unholy traffic. Lamps in Lagos!"[8] However, in addition to new infrastructure, cleaning up the town, and pacifying the interior peoples, the truth behind Glover's secret to maintaining "order" in Lagos was his use of police force and Bridewell, as he referred to the jail (*Ile Ẹwọn* to the local population) in Lagos.[9]

Lagos's merchants and administrators had always been suspicious of missionaries and the improving impacts of missionary activity, preferring to promote the civilizing effects of trade and infrastructure. They felt not only that missionary activity interfered with trade but that efforts to educate Africans, to arm them with literacy, would potentially undermine British authority. They were not entirely wrong; eventually, some of the most compelling challenges to British rule came from the British-educated Saro and the indigenous Lagos elite.

Newspapers and other periodicals began to emerge in Lagos after 1851. Missionaries circulated their news in various journals, and in 1863, Robert Campbell (a Jamaican immigrant) began to publish his *Anglo-African* newspaper. That same year, the African-Aid Society distributed the *African Times,* gathering news from all over the West African coast and publishing letters from correspondents as well.

A flood of letters to the editors of *The Anglo-African* and the *African Times* appeared in the 1860s and 1870s. Their preoccupation? The dispensation of supposed justice in Lagos and the ways it was linked to the administration of the new colony. These letter writers used bright and ostentatious pen names: "Broad-Street Hatch," "Ologun Dudu," "Negro," "Lagosians," "Africa," "Public Opinion," "A Contemplator," "Liars Are Being Found Out," "Consequence," and, eventually in 1872, "The Eko People."[10]

British administrators promoted a public line that they were introducing civilization and peace and maintaining freedom. In their letters and debates, writers to and for *The Anglo-African* and the *African Times* joined local administrators in constructing a new image of the city and its connections across the West African coast and the English-speaking African diaspora, asking, What sort of place was this new colony to be? And for whom was it maintained? These ambitions for a better Lagos, however, came at considerable cost to the Lagosians.

These frequent letters in the press did not go unnoticed, and likely had a detrimental effect on Glover's reputation. There were some positive letters, especially those written by Ologun Dudu, who praised Glover and emphasized just how much the local population appreciated their "Beloved John" or their "Ọba Gọlọba," as he was warmly known, at least by some.

Law and justice were unevenly administered below and above Broad Street, in both a literal and figurative sense. British officials had long considered a range of ideas to secure their interests in the town in the name of civilization. Gunboats were always useful—especially for a show of force—but were easy to evade in the shallow, labyrinthine creeks that surrounded Lagos and connected it to other towns. In early 1861, one consul suggested the use of "black troops, to be styled as a consular guard."[11] In the absence of permanent gunboats offshore, these troops could secure and protect British interests. Officials considered creating a "consular guard" in Lagos to address their concerns over security. Acting consul William McCoskry established the prison and the police in Lagos in 1861. There were originally around twenty-five constables in the police force in 1862, but their beat did not leave the European portions of the town. This meant they clung tightly to Water Street and Broad Street and were sighted occasionally in Olowogbowo. Lagosians complained of robberies in the fields where the police did not visit.

In addition to police rarely venturing above Broad Street, it seems that Black Lagosians received a different application of the law than did White colonials. In their letters to the *African Times,* Lagosians wrote of the "almost indiscriminate handcuffing of men, women, and children, apprehended on the most trivial grounds, and not unfrequently by mere ignorant official caprice, or an unjustifiable exercise of official power."[12] An 1870 editorial in the *African Times* titled "Shameful Mal-administration of Justice at

Lagos" summarized the impact of the most infamous cases of British colonial injustice in Lagos, identifying the police court as a main site of injustice and activism against colonial exploitation along racial and gender lines.[13] It took the arrest of two White Europeans, William Turner and Wilhelm Rascher, to draw attention to the culture of confinement in Lagos, and to make sure changes were finally implemented. This imprudent act on Glover's part—arresting the European men *with* the Africans—cost him allies in the European quarter, if not total respect.

In many ways, the new, "civilized" colonial Lagos seemed not to be built for the ara Eko or ọmọ Eko, but race was not enough to predict the treatment one would receive in Lagos. Several Sierra Leonean returnees worked for the ọba and in the colonial government. To many Lagosians, as commentary following Ọba Dosunmu's attempted rebellion in 1863 demonstrates, these cultural differences mattered perhaps more than the racial similarities. For some Saro, for instance, even though their racial background and ethnic identity meant they were closely linked to the local populations in Lagos and in the lagoon cities beyond, their conversion to Christianity and Western education meant they also had something significant in common with the Europeans. The same was true for the returning men and women from Brazil, Cuba, the United States, and Jamaica.

Antagonism between the local Africans and the British was often clearly defined along racial lines—the peace between the British and the locals was uneasy. In this early colonial state, it was less clear what role the Saro and other immigrants would play as they navigated the spaces between these populations.

Of course, Lagosians had suffered from crimes in their city prior to British rule, from robbery to rape, but without the rhetoric of "civilization." Letters and editorials in the *African Times* complained of the gross abuse of police powers and how, when maltreated, the Africans in Lagos had very little recourse for appeal. These abuses spread from the streets to the seas, where Africans were even maltreated on the ships and steamers that went back and forth along the coast, from Freetown to Lagos. Outside, all along Water Street and Broad Street, Lagosians encountered the strong arm of the law. Inevitably, those Lagosians considered "uncivilized" found themselves at "Bridewell," the infamous nickname for prisons that Glover applied to the jail at Eko.

Annexation meant the power to punish and control had moved south to the Marina. Even Ọba Dosunmu—now the ex-king—chafed under these arrangements.[14] But what was the city that Lagosians wanted, and how had it changed since their ọba had lost power? To answer these questions, I draw from the editorials in the emerging Lagos press, the letters of British consuls, governors, and Liberated Africans returning to the Bight of Benin, and the maps, books, and pamphlets published about this part of West Africa to demonstrate how questions of race and identity in an urban context interacted in determining who was local or foreign, and how that factored into the reconstruction of space. The evidence suggests that Lagos Island was increasingly divided in the 1860s, not by quarter as previously imagined, but by the Broad Street axis.

Invariably, Administrator Glover was at the center of many of the most significant disputes. Far more has been written about Glover's efforts outside the city than those within it; much has been made of administrative efforts to expand and define the boundaries of the Lagos colony and control trade and block roads between Lagos and the Yoruba interior, as well as the stamping out of slavery in the region. In this chapter, I argue that the encounters at the police court made public discourse visible in the city, where newspapers could only capture certain conversations. Eventually, Glover closed the court, but within two years of that action, he himself became a casualty of the events and had to leave Lagos. To say that Captain Glover was a polarizing figure in Lagos is to underestimate the influence of his actions in Lagos, across the lagoons, and in the Yoruba interior.

A WALKING CARTOGRAPHY: MAPPING THE BROAD STREET AXIS

The maps of Lagos—the city and the colony—that survive from 1861 to 1872 make cogent arguments about place, race, and belonging. From the 1871 map of the plots of the Marina to *Lagos et Ses Environs* (see figure 5.1), the men who sketched out Lagos still took the time to maintain the distinctions they observed. Written sources add texture and detail to this spatial argument. In figure 5.1 we see a crude representation of this separation on the island. There is a distinction between the representations of the "Black Town" (labeled 1: "Ville noire") made up of Ọfin and Iga quarters, the "European Quarter" (labeled D: "Quartier Européen") and the Houssa Barracks (labeled 4: "Caimpement des troupes nègres"). Olowogbowo is not labeled, but in this case rests uneasily between the African and European spaces.

The map for this chapter (map 5.2) follows the impact of the Broad Street axis on the political and social life of Lagosians. The physical evidence suggests that this street was created to separate the White and Black populations in the city, and the textual and spatial data concur. The new institutions of British administration span Broad Street and are located at

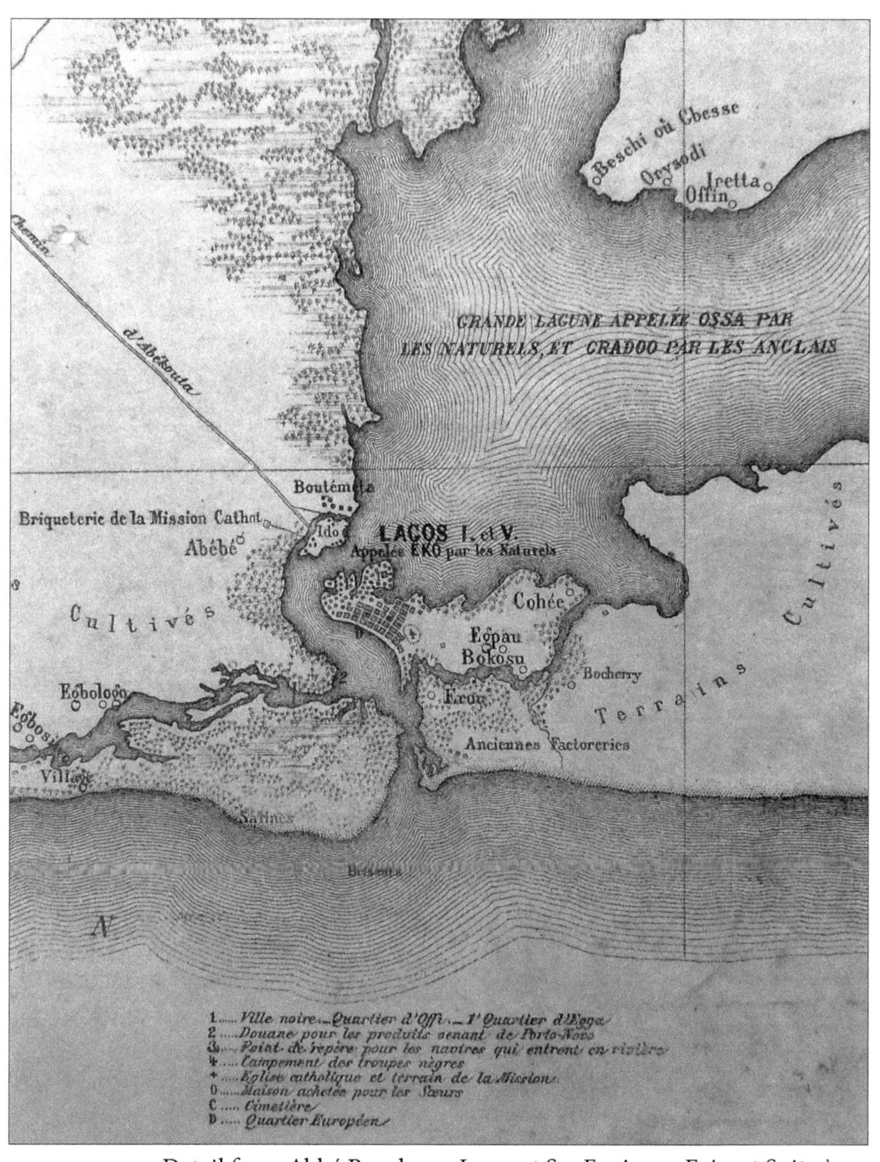

FIGURE 5.1. Detail from Abbé Borghero, *Lagos et Ses Environs: Faisant Suite à La Carte Du Royaume de Porto-Novo (Dans Le Vicariat Apostolique Des Côtes de Bénin),* 1871, 2.5 inches to 1 mile (scale), CO 700/WestAfrica 20/2. Courtesy of the National Archives, Kew, UK.

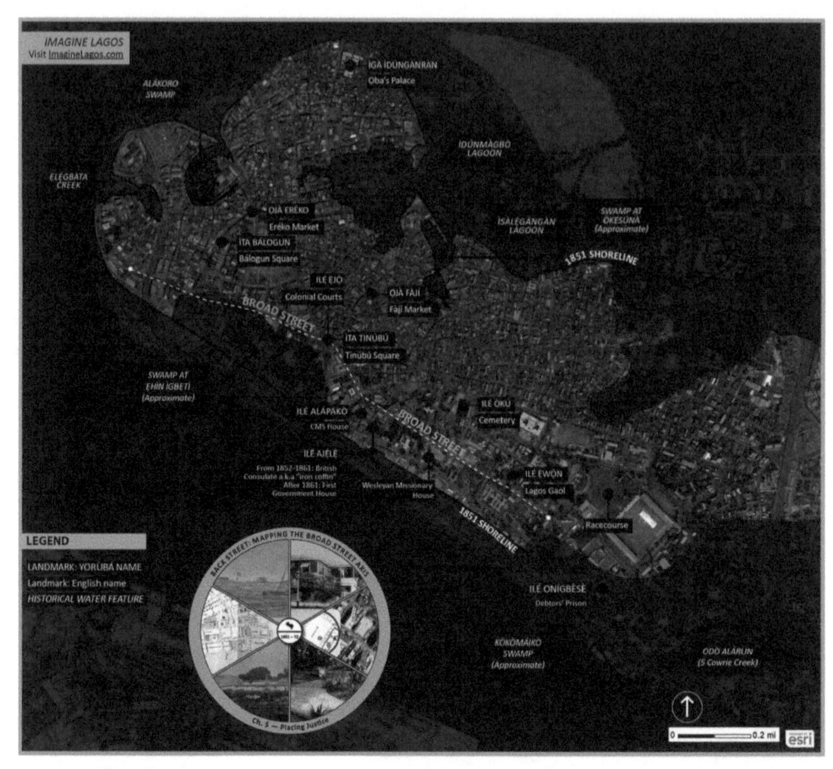

MAP 5.2. "Back Street: Mapping the Broad Street Axis." Base map courtesy of ESRI. Images in the cartouche are, clockwise from top right, as follows: (R1) Government House, Lagos Island, undated, photograph courtesy of the National Archives, Kew, UK; (R2) Spyglass over Prison Street and Broad Street in Lagos Island; (R3) photograph of Freedom Park, former site of the colonial prison, Lagos, 2019, photograph by author; (L3) detail from "The Gaol at Lagos," 1871, CO 147/24, the National Archives UK, courtesy of the National Archives, UK; (L2) detail from W. T. G. Lawson, *Plan of the Town of Lagos, West Coast of Africa: Prepared for Lagos Executive Commissioners of the Colonial and Indian Exhibition 1886; By Order of Fred. Evans, Esq. CMG. Deputy Governor,* 1885, about 1 inch to 275 feet (scale), CO 700/Lagos9/1, courtesy of the National Archives, Kew, UK; (L1) detail from "Lagos, WestKüste von Africa," 1859, reproduced by kind permission of the Syndics of Cambridge University Library. Map by author.

specific intersections: from Tinubu Square, where Customs Street intersects Broad Street and the courts are located, to Government House at Ajẹlẹ Street and Bridewell at Prison Street. Religious institutions like the Church Missionary Society and Wesleyan church play a much-reduced role when compared to the 1850s. Broad Street also functioned as a boundary between the Marina district and the rest of the city.

The images in the cartouche represent sources that mark these spaces in Lagos and abroad. They are (clockwise, starting with the image on the top right) (R1) the "iron coffin" from the 1860s, (R2) a Spyglass over Prison Street and Broad Street, (R3) Freedom Park, Lagos (formerly the colonial prison), (L3) the Lagos jail in 1871, (L2) W. T. G. Lawson's *Plan of the Town and Island of Lagos,* 1886, and (L1) "Lagos, Westküste von Africa," 1859. The fix-and-fill maps are most useful here in making this map, as they retain some idea of ownership of spaces.

RACE, PLACE, AND INJUSTICE IN LAGOS

In the 1860s, both the *African Times* and *The Anglo-African* published stories on controversial political and economic activities and incidents, including the 1863 case of James Cole, the 1868 case of Thomas Lubley, and the 1869 case of William Turner and Wilhelm Rascher. These accounts provide important contrasts to those found in Colonial Office documents that include confidential administrative letters written from Government House on the Marina to the Colonial Office in London, which have formed the basis for such historical treatments as that of Spencer Brown, or missionary periodicals like the *Church Missionary Intelligencer* and *Wesleyan Missionary Magazine,* which were focused on promoting Christianity and the work of missions in Africa.

Although many writers for the *African Times* and *The Anglo-African* were unidentified, it is from these writers that we learn of some context behind important events in Lagos. In one interesting case, an unnamed writer recounted Ọṣodi Tapa's position, death, and interment. In the letter, they acknowledged his importance, not just to Ọba Kosọkọ but also to John Glover. To the writer it explained why Tapa's burial had special circumstances: he did not have to be buried in the public cemetery, and, in fact, the government bore the cost of the fashioning of a vault for him on Ọṣodi Street in Ẹpẹtẹdo. Glover and Ọṣodi's closeness was even reflected in the city's layout: their eponymous streets were parallel and adjacent.

The Cole Affair (1863)

Local women were not always safe from the advances of European men, and it seemed there was no place where they were totally protected.[15] Late in 1863, a young woman alleged that she had been assaulted by two European men on horseback while she was fetching water from a public well. Their assault then extended to James Cole, a Sierra Leonean man whose compound she ran into for help. In court, Cole testified that it was while trying to help the girl that he was dragged into the matter. He knew her and she was described as "part of Rev. Crowther's household."[16] That she was from Crowther's compound and ran into Cole's home meant she was most likely in the Olowogbowo area, at the public well that had recently been sunk at Balogun Square.

There are two surviving accounts of this episode. In 1863 Cole claimed that when the girl ran into his compound, chased by two Europeans on horseback, he tried at once to intervene, asking, "What's the matter?—any harm sir?"[17] Their response was to flog him with a whip. He tried to remove them from his premises, but they continued to whip him. These events were interpreted in terms of race and morality—the former relating to the impudence of the European men, and the latter on the so-called improved morality that the Europeans were supposed to bring to Africa. When considered from the position of space, we see how Blacks—be they immigrant or native—were still not safe in their own quarters.

Cole made immediate complaints to the police but was rebuffed and told to "settle the matter privately."[18] He persisted, and the magistrate's court decided that the two men should pay a fee equivalent to the amount of the summons: three shillings. In March 1864, a letter to the editor of the *African Times* offered a different take on the matter. Writing as "One of the defendants" (pseudonym), the author emphatically denied Cole's description of the case, claiming that he and his companion did not attack the girl, and certainly not Cole either.[19]

He claimed he knew the girl and was actually trying to make conversation with her in a way that was neither lascivious nor aggressive, unlike what was implied by Cole's testimony. He claimed that he never touched Cole either, and only resorted to threatening to whip him when he became exasperated after a scuffle ensued while trying to extricate themselves from Cole's compound. Tellingly, he gave no concrete explanation for trespassing on Cole's property, or what the case would have been if role, race, and gender had been reversed. The injustice from this case of a public attack on a Sierra Leonean lingered in the minds of Lagosians, and their letters to the *African Times* raised the issue again, seven years later.

The Lubley Case (1868)

An author under the name "Broad-Street Hatch" wrote a letter titled "How Justice Is Done at Lagos," which highlighted one of the worst cases of child sexual abuse recorded at the time.[20] In 1868, a young Black girl of around seven or eight was seen running out of the home of Thomas Lubley, a European "gentleman bookkeeper" for the firm of Kidd and McCoskry, in

Olowogbowo, and found bleeding.[21] She needed significant medical attention from the assault and was said to have bled for four days afterward. Some people feared she would die from her injuries.

Lubley pleaded not guilty in the police court, but the jury found him guilty of rape and sexual assault. The whole town awaited his punishment, but the magistrate sentenced him to only four months. Hatch (pseudonym) wrote that the Lagos community—at least the Africans—was outraged by this sentence. "We demand equal justice," wrote Hatch; he knew a Black defendant would have been sentenced more harshly.[22] Lagosians were acutely aware of the differences in treatment and privilege of the White and Black populations in Africa and all over the coast. From The Gambia to Freetown and the Gold Coast, the *African Times* carried stories of the growing disillusionment with colonial European rule. "Our eyes," wrote an author under the name "Lagos," "are on him, the court, and the judge."[23]

Benjamin Way was a controversial magistrate and, in Palma, had previously sentenced an African man accused of rape (in a less serious case) to three years of hard labor.[24] Lubley was said to have all kinds of privileges while in jail, while the typical African, even when not "tried, convicted or condemned," was denied proper food, clothing, adequate shelter, and even access to family visitors.[25]

Justice (pseudonym), writing in 1869, claimed to be one of the jurors in the case. The case brought up tensions around sexual politics in Lagos, especially in the relationships that European men had with African women. Lagos (pseudonym) wrote of the hypocrisy of European residents, especially in terms of spreading "civilization" to the Africans. In "European Example in Africa," he wrote of how European men interacted with African women. "One," he wrote, "lived in open decency with three loose unmarried girls." While others practiced the polygamy they condemned the Africans for, having mixed-race children out of wedlock with different African women, and consorting with "abandoned women" (likely a euphemism for sex workers or women who had culturally fallen out of favor). As for the Lubley case, Lagos wrote that there would be "no consequence perhaps. Why? Because he is a white man."[26]

The Rascher and Turner Case (1869)

The matter of policing Lagos was fraught. The arrest of two White European men in 1869 illustrated the conditions of the prison in Lagos, and the ways Administrator Glover acted with impunity. If these two White men could be treated so badly, what did it say about the treatment the local population endured? This was exactly the argument made by the letter writers in the *African Times.* The arrests of William Turner, Captain Wilhelm Rascher, and C. F. Meyer's Kroomen illustrated the deplorable condition of the Lagos jail and highlighted the terrible treatment meted out to locals. Glover produced a sketch map of the eastern section of Broad Street in his defense to his superiors.

Glover's confidential correspondence regarding the case showed the private contempt he held for some of the African population, even though publicly he was praised as "Baba Gọlọba." Glover's 1869 sketch of the Marina shows Government House at the intersection of Marina Street and Ajẹlẹ Street, with Meyer's compound right across. With this diagram, he attempted to show the proximity of the Kroomen, whom one of his administrators described as "thirty or forty savages, howling."[27]

Turner testified the Kroomen were simply celebrating with some gin that their boss, Captain Meyer, had provided for them that evening. Turner and Rascher were smoking in the yard, and the men had wound down. Suddenly, forty Hausa men bearing cudgels and arms burst into the yard, attacking the men. The Europeans and Kroomen were arrested. Glover came by to inspect them and was said to have declared, "Handcuff both these men and take them to Bridewell."[28] The Lagos prison was named as a reference to a notorious prison in London, which allows us to understand the state of the place at Lagos. Turner and Rascher were handcuffed to each other, marched two miles through the streets of Lagos to Government House, and finally deposited at the jail. Since the distance from their house to the jail was less than half a mile, this display was likely organized to humiliate them in the town. Testis (pseudonym), writing about the "gross outrage on Mr. Turner," maintained that the so-called police were but "ragamuffins" and "ruffians" who operated according to Glover's caprices.[29] Policing in Lagos seemed to mean merely doing "Captain"—used in the letter derisively—Glover's bidding.

Turner and Rascher were initially imprisoned outside in the yard with the Black *ẹlẹwọn*—inmates—but protested that as White Europeans, they had to be kept indoors and not exposed to the elements. In his affidavit, Turner described the insides of the jail as "stinking with excrement and otherwise in a most filthy state."[30] They decided to stay outside. In their case, they were lucky and were released within a few hours. Turner said he was back in his home by 2 a.m. An 1869 editorial in the *African Times* complained of the gross abuse of police powers, and how, when maltreated, the local natives had very little recourse to appeals. The editorial noted that there was an "almost indiscriminate handcuffing of men, women, and children, apprehended on the most trivial grounds, and not unfrequently by mere ignorant official caprice, or an unjustifiable exercise of official power."[31] The editorial noted that this case was an interesting turning point for the European community, which was usually indifferent to the abuse of the local population and, in fact, sometimes even party to the misconduct. This time, however, Administrator Glover had used his "quarter deck instincts"—a reference alluding to this former naval captain's snobbery—to treat Europeans as he did Africans.[32]

I'LL SHOOT YOU AND PAY 10 POUNDS: THE WRECK OF THE *THOMAS BAZLEY*

Did Lagosians sabotage the steamer *Thomas Bazley* in protest of racism and colonial malpractice? Or did it sink and "became a total wreck" because of an unfortunate accident at the treacherous sandbar?[33] In November 1869, the ship washed up mysteriously on the shores of Lagos. This shipwreck marked a controversial end to an episode that had stirred up a lot of turmoil around race, masculinity, place, and the administration of justice in Lagos and along the coast and, eventually, landed the ship's master and one Black passenger in the pages of the *African Times.* On November 16, 1869, Public Opinion (pseudonym) wrote an alarming letter to the editor of the *African Times.* It began, "The great cry in Lagos now is I'll shoot you and pay 10 pounds," alluding to the matter of Captain George Stott's October 1869 abuse of Isaac Baker aboard the *Thomas Bazley.*[34] Baker was a Black retired soldier from the Second West Indian Regiment now working as a seaman. He went up the Niger River on the steamer and came down on the same boat in October. Stott was the master of the ship. Baker was in agony in the face of the abuse but was optimistic that once they arrived in Lagos, he could get some modicum of justice for being ill-treated by the White captain. Though the incidents happened on the ship, they had an alarming effect on the Lagos population, as the case was tried in the Supreme Court in the city.

The *Thomas Bazley* was only one of many controversies in the city in the first decade of formal colonial rule, and the incident was reminiscent of the ill treatment that Black Lagosians faced constantly: from unprovoked whippings to sexual assaults and abandonment. The new colonial regime seemed to do little to remedy the various plights of the ara Eko. It was only when these assaults happened in public, either in the streets or in the compounds, that it was possible to even seek justice.

At least three different letter writers addressed Captain Stott's attacks on Isaac Baker on the *Thomas Bazley*. Each writer gave a version of the horrors that Stott inflicted on Baker. Consequence (pseudonym) wrote of the "horrible brutality" aboard the steamer; of the "gross injustices" that the "colored population" faced in the face of European rule, simply, as they said, "because of their color";[35] and of how local Europeans thought it foolish for the local Africans to appeal for justice, especially as they were uncivilized. Stott was accused of inflicting a series of punishments and horrors on Baker: from whipping him, to locking him in the ship's hold, to starving him, and to trying to "kick, beat and stamp upon him."[36] He also threatened to end Baker's life by shooting him with a revolver, and heaped upon him the most horrific verbal abuse. In the account by Public Opinion, in addition to shooting at Baker, Stott also handcuffed the man and "triced him up like a beast so that his toes just touched the deck."[37] Baker himself alleged that Captain Stott had tried to murder him twice. Later, when Stott was questioned about his treatment of Baker, his response was inflammatory but not surprising to Lagosians who knew of the case: "The law was never made for black niggers."[38]

The trial was held on November 3 at the courthouse in Tinubu Square. Justice pointed out that this case was particularly bad because it was reminiscent of the Lubley case of 1868, where another serious miscarriage of justice occurred. That case had had a grand jury (eight Africans and four Europeans) and then a petty jury (made up of eight Africans and two Europeans). They had recommended a three-year sentence. In the Stott trial, Stott pleaded not guilty, and the jury was split against racial lines, with two Europeans against ten Africans. The magistrate sided with the Europeans that Stott was not guilty. Consequence claimed they (the jurymen and Stott) were "mason brothers" but perhaps their shared race was enough to spoil the vote. There was an immediate outcry in Lagos in protest of the verdict. The jury was "locked up" and when the magistrate visited at 7 p.m., they were still deliberating and could not agree; he let them go. Benjamin Way, the magistrate (siding with the two Europeans), instead imposed only a paltry fine of ten pounds. Consequence wrote that it was a "heinous" case, and "everyone [was] disgusted at the glaring partiality shown in one of our so-called courts of law." They alleged that Baker was offered a seventy-pound bribe to "not appear in court" and added that there was a sense in the city that "there are so many underhand influences at work to shield criminal Europeans."[39] "Such is the state of our courts of justice," Consequence lamented.[40] In response, Glover suspended the magistrate and closed the court until February 1870.

Undeterred by the outcry, Stott and his fellow European masons celebrated publicly for three days at the premises of the West African Company on the Marina. One evening, they chose the home of William Reffle, a Liberated African in Lagos. Reffle's property in Lagos Island was a small triangular plot north of Water Street. Reffle had allowed the use of his property by a European friend, but he wrote to the *African Times* that he had not been made aware that the dinner would be in honor of Way, the departing chief magistrate. Reffle wrote that in his work on two cases—that of Lubley and of Captain Stott—the magistrate "endeavored to display his zeal in the interest of his countrymen rather than the execution of the law."[41] Reffle had signed a petition against the actions of the magistrate and was concerned that the party on his premises (in his new building) would encourage people in Lagos to believe that he had changed his mind and was now in support of the magistrate and the White masons.[42]

In protest of the verdict and light sentence, the entire crew of the *Thomas Bazley* refused to serve. They were discharged. Three days later, Stott assembled a small crew of six Kroomen, an engineer, a cook, and a firefighter to get the ship over the bar. Consequence claimed that it was because it was "not being properly maintained by seamen [that it] met an awful fate, being wrecked to pieces on the bar." No one died.[43] Consequence saw the wreck as a final judgment from God, a new karma wrought by the sandbar. They wrote that "Stott slipped out of Lagos" after claiming that the ship was well manned.[44]

ỌBA GỌLỌBA

> Whatever is written to cannot be worse than the fact. Captain Glover has much to contend with against Europeans, because he loves the African and they hate Africans. May God Bless you and spare your life for the good of Africa and Africans.—yours truly, "Public Opinion."[45]
>
> —Letter to the editor of the *African Times*

These injustices on the Lagos population were at odds with the public face of John Hawley Glover, the governor. Glover was a sort of champion for the local population, at least by those he considered allies. The debates about Glover point to the differences between public and private discourse and the difficulty of drawing from sources designed specifically for certain audiences. In the case of the confidential letters, Glover and other administrators wrote accounts that were beneficial to them. And they were not open to the scrutiny of the public that they lived with. Administrators like the secretary of state were far from these colonies and received these letters at a clear remove. Somewhere in the gap between the material that was confidential and that which was public is the way that these events may or may not have left their imprint in space, and how they shaped place.

By the time Glover arrived in West Africa, he already had colonial experience in Myanmar and China. He was in his late twenties when he joined William Balfour Baikie's second voyage on the Niger River on the *Dayspring* in 1857.[46] He was named Commander Glover of the gunboat *Handy* in 1862; he was made a captain (a courtesy title) in 1877, after his departure from Lagos.[47] The *Dayspring* hit a rock after they left Jebba, and the team camped overnight. A tornado struck and wrecked the ship; it had sunk by the morning. Glover was noted as the first European to survey the Niger River. It was during his return trip over land that he gained more experience with Hausa porters and different Yoruba groups, including the Ẹgba and Abẹokuta; and these experiences were said to have shaped his actions at Lagos.[48]

Back in Lagos, he occupied himself with surveying the lagoons while he waited to return to the Niger.[49] He took the overland route to Lagos, from where he got a ship back to Sierra Leone.[50] W. D. McIntyre noted that based on Glover's own experiences in the interior, Glover interpreted Lagos as a sort of headquarters for more action in the interior. He returned to Lagos in March 1861. McIntyre points to four problems in Lagos: domestic slavery; ill-defined boundaries of the colony; French competition, especially at Porto Novo; and the warfare raging in the Yoruba interior. Glover was made administrator in 1866.[51]

Letter writers viewed the *African Times* as a venue to share their perspectives, even those that countered the colonial government. In one 1871 letter to the *African Times,* the writer noted, "We are glad to notice that your valuable paper is open to both high and low, rich and poor, who delight in truth."[52] The truth about the administrator's reputation in the city was hotly debated. However, it was not until his actions provoked an economic crisis in the city that he was replaced as administrator. In 1872, the crises in Lagos reached a head: the roads were closed, trade was virtually nil, and there was great turbulence in the town. Grumbling against Glover increased.

Historians have covered the implications of Glover's actions in the Yoruba interior, and the effects on Ẹpẹ, Porto Novo, Ibadan, Ijaiye, and Oke Odan.[53] Two separate communications (writing for and against Glover) attempted to speak for the ara Eko and ọmọ Eko in the months before Glover's departure. Glover's allies in Lagos included Sierra Leonian clerks who worked with him in his administration. He also formed alliances with prominent African traders like Oṣodi Tapa and Taiwo Olowo. One official noted in the minutes that the "pure native portion" of the town supported Glover, especially in the building of roads.[54] In one case, Taiwo Olowo (the well-known merchant of Taiwo Street) arranged a letter in support of Glover. He and Ọba Kosọkọ addressed the letter to Glover.[55] Officials expressed doubts in the minutes attached to the letter, especially because of the similarity of the letter signers' handwriting and the uncertainty of Glover's actual intentions. This letter caused a stir in Lagos, and different parties had ideas about who the actual writers of the letter might be.

In another letter signed by the "aboriginal natives of Lagos" they defended Glover from the attacks printed in the *African Times*.[56] They argued they were initially against the cession, but were now fully in support of the British government's actions at Lagos and beyond. They alleged that the critical *African Times* letter writers were likely the "civilized Ẹgbá" at Lagos ("both European and Native") and they cast the Ẹgbas and Ijẹbus as "only obstructors" to free trade.[57] They referred to letters from 1871 and 1872, likely the ones written by "Africa," "Vox Populi" and "Observer."

The "aboriginal natives of Lagos" claimed to speak on behalf of the ọmọ Eko, and of the "whole country," and appealed to Lord Kimberly (who was then colonial secretary) that Glover should not be removed from Lagos, even if he could find another suitable administrator. They argued that the *African Times* had predicted that Glover would be removed, but that they should reconsider, a move that would be "ruinous." The signatures were led by Kosọkọ and Taiwo; Ọba Dosunmu was conspicuously absent. Fifty women and 120 Lagosians signed the letter. Men and women were listed separately, but each had an "X" by their name. The names offered a veritable tour of the indigenous parts of Lagos as several men indicated which part of Eko they were from. Signatories included Esubi and Tokosi of Idunmagbo, Brimah of Alapatira, Ayorinde of Idunmọta, Sibu of Ita Obadino, Bakary Ikoni of Oke Popo, Fasinnah of Ita Faji, Abason of Oko Awo, and Agbojo of Idoluwo, among others. These public letters are often the only times that these historical actors entered the record, and this is one of the first times they ascribed their locations to their names. Indigenous places were legible now in Lagos.[58]

Glover's problems were mounting. In 1871, Pope Hennessy, governor-general on the west coast, wrote, "The Prison at Lagos is a disgrace," echoing concerns raised in the case of Turner and Rascher.[59] He wrote again in 1872 that Glover had been told for years to repair the prison.[60] The jail at Lagos was in disrepair, but Glover disregarded numerous requests to fix it.

FIGURES 5.2A–C. Photographs of the debtors' prison, stables, and jail in Lagos, December 1871. Courtesy of the National Archives, Kew, UK.

Since 1865, Glover had promised to address the problems in the jail, but never did, preferring instead to focus on his stables and improvements at Government House. Pope Hennessy ordered photographs be taken of the public buildings in Lagos (see figures 5.2a–c). Four have survived: the debtors' prison on the Marina, Glover's stables, the Lagos prison, and the jail at Badagry. All, except the stables, were in bad shape, with their thatch roofs failing and their compounds decrepit.

Despite his public detractors, Glover was not without defenders. Ologun Dudu (pseudonym) wrote at least two letters in 1869 in his support. In one letter titled "Good Governance at Lagos," he praised the administrator for the state of the city, writing, "Lagos is much improved, and is improving daily."[61] Some of these letters expressed clear anti-Yoruba sentiments, declaring that Yoruba people thought themselves better than other ethnic groups in the town. "They are a band of idlers," wrote Africa (pseudonym), "consisting of Sierra Leone and Brazilian refuse, of pagan, Mohamedans [*sic*] and polygamists."[62] Balogun Street and Government House were said to be their favorite haunts, and Africa accused them of "sowing discord" in Lagos. All these writers remained anonymous in public, but there are clues in the letter of who each writer suspected the other writers to be. Perhaps the most obvious would be Ologun Dudu, who, as a literate African so partial to colonial government, was likely one of a handful of men, from James Davies to Otonba John Payne, who at the time were on friendly terms with the British administrators. Further, Ologun Dudu wrote specifically of the street names and their significance, praising Glover for honoring the "old men in the town," prompting suspicion that he might have been Otonba John Payne.

News of this letter spread around town, but it was already on the way in the mail steamer to the secretary of state. The Eko People (pseudonym), writing in the *African Times*, accused Taiwo of getting "about eighty ragamuffins to put their marks in the petition, against their own conscience."[63] This writer accused Taiwo of plotting revenge against those who did not sign the letter and, in fact, said Taiwo marked an "X" for names he listed. Claiming to speak on behalf of Lagosians, the writer added, "We have nothing to do with it."[64]

However, Glover was sent away from Lagos in the summer of 1872. Writing that same month, someone under the name "Liars Are Being Found Out" noted, "Our fullest hopes have been answered in the departure of Captain Glover from the Shore of Lagos."[65]

One writer was particularly disgruntled. Writing on November 1, 1871, Negro (pseudonym) attempted to discredit nearly all the writers quoted here. In 1871, they took aim at "Olitoh" for writing "all rot," especially in their vocal support of Administrator Glover. They accused the editor of the paper of having "allowed the friends of Glover to tell you the bright part of Lagos always."[66] Writing in 1871, Africa (pseudonym) offered that it was their "duty to tell the world the dark side of it." In their telling, Olitoh (pseudonym) was clearly a Glover partisan, and was therefore distorting what was unfolding at Lagos. However, as they wrote presciently, "You will soon see the handwriting on the wall, 'Mene mene, tekel upharsin,' at Government House in Lagos."[67] Ultimately, Africa was proved right: Pope Hennessy was dissatisfied with Glover's conduct. In June 1872, Glover took a leave of absence. He never returned to Lagos. A quote from a minute in 1873 articulated this turning point in old Lagos:

> Whatever may be the policy of this or any other Government, Public opinion will not permit the withdrawal of British Authority from the W. Coast of Africa. Then what is to be done? Why did we ever occupy Lagos? To put down slavery and to extend Christian civilization. Are these still considered objects worthy of your attention? If not, the sooner the withdrawal the better, and if English opinion will endure our withdrawal no doubt a certain amount of expenditure will be saved and we shall be spared some trouble. But if these are still objects to be aimed at, and if, as I think, a great country which has undertaken certain responsibilities of this kind cannot evade or abandon them without loss of Honour and Character, then surely some definite course of action should be adopted.[68]

~

Why did the new administration choose Ita Tinubu as the site of its primary intuitions and, in doing so, keep the name of the wealthiest female merchant, whom they had exiled? It could be interpreted as one of many acts of rewriting and reorienting space in old Lagos. As Tinubu Square, the site had been linked to past regimes marked by indigenous control and slave trading. By placing these new institutions—the petty debt, criminal, police, and slave court—there, the administration was reclaiming space in the new heart of the city. But this new regime did not make any promises that it was for the betterment of Lagosians.

Yet, looking back, Ọṣodi Tapa's death on July 2, 1868, was the harbinger for the end of an era.[69] Glover's greatest local ally, Ọba Kosọkọ, passed away on April 26, 1872, barely two months after endorsing the letter in support of his administration. In the end, it was force and not indigeneity that prevailed in establishing ownership over the entire island, and once it had crushed Dosunmu's initial rebellion, the new administration focused on the task of securing property to ensure that the new colony would prosper and, eventually, pay for itself. Both local and foreign alike were absorbed under the jurisdiction of the British, who demonstrated that their naval power made up for whatever claims they lacked in terms of the right to the island and port.

Conclusion

Eko o ni Ranti?

> Lagos exerts a secretive, sometimes resented, but tenacious hold on all who pass through its steamy streets and tumultuous markets.
>
> —Wole Soyinka, "Unsinkable City"

In the decades before the almost synchronized independence of the 1960s, while Africans migrated to cities in increasing numbers, Africa's historians were still mostly preoccupied with rural and peasant studies. Africa was still envisaged as an "entirely rural world,"[1] where almost all instances of urbanization or city life were attributed to external forces, be they European, via colonization, or because of Islamic influence. Historical research on cities gained momentum in the 1970s, but this was after a large, existing body of urban literature was already involved in a discourse focusing on how to define a city, without specific reference to Africa.

Lagos of the early 2000s—freshly in the news as dark, dangerous, and dirty—seemed in free fall at the beginning of the new millennium, despite the promise of a newly elected democratic government. As a researcher and editor at the architect Rem Koolhaas's Office for Metropolitan Architecture (OMA) in the early to mid-2000s, I did research on Lagos's post–oil boom infrastructure. This work made it clear that the roots of rapid urbanization, the oil boom urbanism, and the consequences of a militarized state had to be understood from the perspective of a longer time span. Since then, I have channeled my background in architecture and engineering to understanding the city's intertwined histories.

MAP 6.1. Spyglass over Ita Tinubu / Tinubu Square in Lagos Island. Base map and Spyglass courtesy of ESRI. The historical map is a detail from *Plan of the Town of Lagos—West Africa (in 15 Sheets),* 1891, 1 inch to 88 feet (scale), 63 × 90 cm, CO 700/Lagos 14, courtesy of the National Archives, Kew, UK. Map by author.

Now, Africa's cities are among the fastest growing in the world, and the teeming populations, inequitable distribution of resources, and struggling infrastructure are a volatile mix even at the best of times. Recent scholarly and popular attention has helped to recognize the importance of Lagos, this burgeoning city of approximately twenty million people. This book addresses the city's nineteenth-century history, which, as I show, is a period that laid the foundation for Lagos's current growth and its political, cultural, and economic preeminence in the region. As such, urban questions have taken on a new urgency in Africa as cities like Lagos continue to grow faster and more intensely, seemingly every day. Most often, engagement with these cities is framed in terms of their problems: too crowded, too disorganized, too frenetic. But how should we understand the ways the past shaped and produced this present?

Urban spaces are neither "inert" nor "blank," and the work of historians like Rashauna Johnson has demonstrated how "people produce places" through their quotidian interaction with their environment and each other.[2] Regional forms of power were inscribed in Lagos, from the Bini influence in the settling of Isalẹ Eko to the inscriptive power that slavery had on the city's shape and uses. Guadalupe Garcia, writing about Havana, has pointed to how colonialism had a specific "physical component in which spatial relationships were central to the exercise and proliferation of colonial power."[3] In this case, there were tangible layers to how British power was, at first, destructive, with the bombardment of the city in 1851; speculative, through the consular period from 1852; and then practiced as permanent, with the annexation of the city in August 1861.

My approach has been multipronged: as a historian, I detailed and analyzed the city's layered past, and revealed how Lagos has changed through time. As a digital humanist and mapmaker, I visualized these historical changes in ways that not only make the complicated history of Lagos more legible but also showcase the contemporary resonances and urgency of that past.

Writing the history of a city in Africa (outside of the north) remains largely a twentieth-century affair. As these cities grow faster and more fiercely than other places, with population often outpacing infrastructure and services, scholars have tried to understand the resilience of the populations who struggle to makes cities like Lagos, Johannesburg, and Kigali their home. Slum clearance, the limitations of the informal economy, and "resourcefulness without resources" frame the importance of these issues, with scholars like Abdul Malik Simone,[4] Achille Mbembe,[5] and Ato Quayson[6] at the forefront of reimagining what it means to be urban in Africa.

Africa's urban historians have argued that there is no single urban history of Africa, as there is no "single paradigm" that can explain the cycles of "growth, development or decline" of these cities.[7] However, too many projects on urban research have the premise that urban change began most significantly with colonial rule. But what spatial narratives did colonial rule produce, remove, and then write over? And how do cities like Lagos remember and mark their own indigenous past?

In the poem "Eko," the poet describes the city as "the untold story of today, webbed in yesterday, hotly at the heels of tomorrow."[8] I have argued in this book that we cannot imagine Lagos without understanding the Lagosians' spatial relationship with their city and island. The book began with the streets and the narratives embedded within. The street names told individual and collective stories, and some still resonate in the contemporary city. It also ended in these same streets, reflecting on the myriad ways the early years of British colonialism changed the direction of so many fortunes in the city.

While some cities have figured out how to live alongside the past, twenty-first-century Lagos Island seems to be conspiring against itself in terms of preserving and archiving its history. Lagos Island feels like it is going through an accelerated and somewhat deliberate scheme of forgetting its past (as referenced in the title, which translates as "Lagos will not remember?"). But there are some bright spots against this tidal wave of

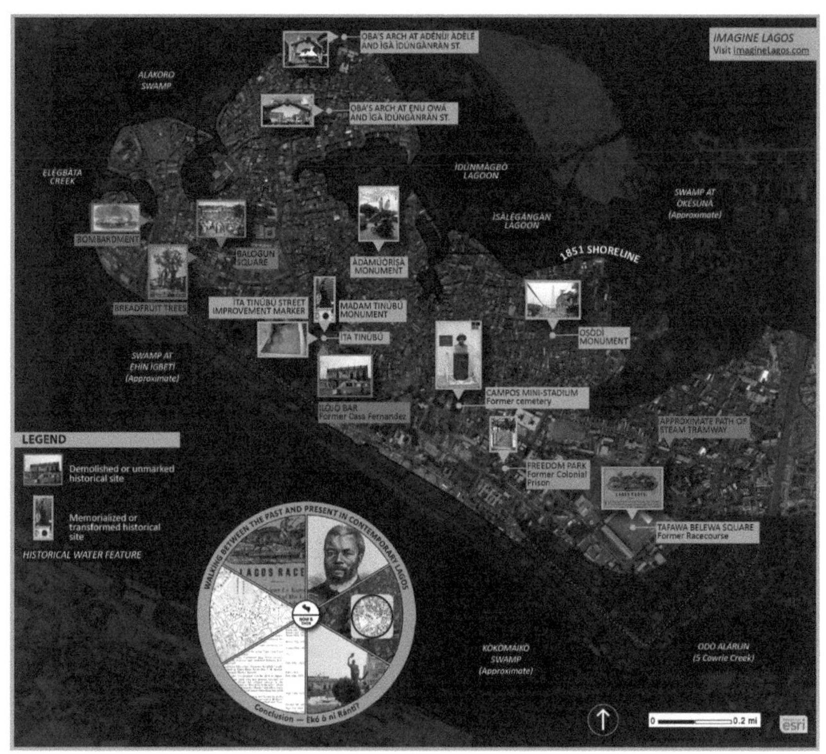

MAP 6.2. "Walking between the Past and Present in Contemporary Lagos." Base map courtesy of ESRI. Images in the cartouche are, clockwise from top right, as follows: (R1) portrait of Otonba John Payne, in Payne's *Lagos and West African Almanack and Diary* (London: J. S. Phillips, 1874); (R2) Spyglass over Tinubu Square; (R3) Madam Tinubu's statue in Tinubu Square, with Ilọjọ Bar (formerly Casa Fernandez) in the background, 2011, photograph by author; (L3) Lagos timeline, in Payne's *Lagos and West African Almanack and Diary;* (L2) detail from W. T. G. Lawson, *Plan of the Town of Lagos, West Coast of Africa: Prepared for Lagos Executive Commissioners of the Colonial and Indian Exhibition 1886; By Order of Fred. Evans, Esq. CMG. Deputy Governor,* 1885, about 1 inch to 275 feet (scale), CO 700/Lagos9/1, courtesy of the National Archives, Kew, UK; (L1) Lagos races in *The Anglo-African* newspaper, 1863, courtesy of the National Archives, Kew, UK. Map by author.

FIGURE 6.1. Madam Tinubu's statue in Tinubu Square, with Ilọjọ Bar (formerly Casa Fernandez) in the background, 2011. Photograph by author.

forgetting, anchored mostly in the work of heritage groups and descendants of the original settlers in ọmọ Eko, who want to preserve their cultural and architectural legacies, despite all odds.[9]

Map 6.2 references a range of commemorative efforts that dot the Lagos Island landscape, marking people and places with complicated legacies (such as the statue of Madam Tinubu, the female entrepreneur and erstwhile slave trader; see figure 6.1). Some landmarks, like Ilọjọ Bar (formerly in Ita Tinubu), were destroyed and not replaced; other sites were deliberately written over with varying levels of success—for example, the Old Burial Ground, racecourse, and colonial prison became a mini-stadium, Tafawa Balewa Square, and Freedom Park, respectively.

Since the turn of the twentieth century, extensive land-filling projects have made Lagos Island nearly unrecognizable. Lagos is still being actively remade, despite all odds. Yet, the old parts of the city are experiencing a curious resurgence based on the reinvention of colonial landmarks. Some familiar nineteenth-century landmarks and roads still frame the western side of the island. The racecourse has been reimagined as Tafawa Balewa Square, where Nigeria's independence from the British was celebrated in October 1960. The Broad Street prison has been recast as "Freedom Park," part cultural center and part historical landmark in the city.

Several themes in the book still resonate in contemporary Lagos. From the entanglements in the land grants to the complex issues around identity and indigeneity, this book shows how deeply the city's nineteenth-century history informs its present. Lagos now is more expansive than just the western side of Lagos Island; the mainland, beginning with Ebute Mẹta and stretching north past Ikẹja, swells with new and old populations looking to carve a path in this city and claim a space for themselves. Now, some unofficial estimates peg the population at close to an unimaginable twenty-two million people, demonstrating the exponential growth of the place and its people.

Perhaps the only constant in the tumultuous history of Lagos has been the sense of ambiguity that has haunted indigenes, residents, and visitors alike. There is a similarity in their letters, newspaper articles, and drawings about Lagos: as challenging as it is to stay in Lagos, it sometimes feels impossible to leave. Being away from Lagos has provoked a curious nostalgia tinged with fear and regret, but the pull is always there. This historical feeling is what is referenced by Wole Soyinka in this conclusion's epigraph, where he identifies the pull of the space's energy, but also the elements that push so many out.

By arguing about historical encounters in space, this book frames the city's past as one open to possibilities rather than bound to the inevitabilities of power and place. However, the historical record often flattens the complexity of this feeling. Lagos is often only one thing: a city that is *leavable* but not *livable*. *Imagine Lagos* attempts to return some of this ambivalence to the city's historical record.

Notes

INTRODUCTION

1. The Royal Geographical Society in Kensington and the National Archives in Kew house the largest collection of maps of Lagos. The British Library, Cambridge University Library, the Bodleian Library at Oxford University, the Nigerian National Museum in Onikan, Lagos, and the National Archives in Ibadan have a selection as well.
2. Ademide Adelusi-Adeluyi, "Historical Tours of 'New' Lagos: Performance, Place Making, and Cartography in the 1880s," *Comparative Studies of South Asia, Africa and the Middle East* 38, no. 3 (2018): 443–54.
3. "Carte de Côte Des Esclaves: Dressée Par F. Borghero, Missionaire, En 1865," 1865, FO 925/454, The National Archives, Kew, UK.
4. Binyavanga Wainaina, "Wangechi Mutu Wonders Why Butterfly Wings Leave Powder on the Fingers, There Was a Coup Today in Kenya," *Jalada Africa* (blog), October 17, 2014, https://jaladaafrica.org/2014/10/17/wangechi-mutu-wonders-why-butterfly-wings-leave-powder-on-the-fingers-there-was-a-coup-today-in-kenya-by-binyavanga-wainaina/.
5. John Adams, *Sketches Taken during Ten Voyages to Africa between the Years 1786 and 1800* [. . .] (London: Hurst, Robinson and Co.; Constable and Co. etc., 1822), 23.
6. Richard F. Burton, "A Day at Lagos," in *Wanderings in West Africa from Liverpool to Fernando Po: By a F. R. G. S. with Map and Illustration* (London: Tinsley Brothers, 1863), 2:171.
7. David Morton, *Age of Concrete: Housing and the Shape of Aspiration in the Capital of Mozambique* (Athens: Ohio University Press, 2019), 12.
8. Vincent Brown, "Mapping a Slave Revolt: Visualizing Spatial History through the Archives of Slavery," *Social Text* 33, no. 4 (December 1, 2015): 138.
9. For their full complaint, see "Banner Brothers to the Right Honourable Earl of Kimberly," November 30, 1870, CO 147/19, The National Archives, Kew, UK; "Copy of Original Title Deed of Banner Brothers and Co's Property at Lagos: Inclosure in Banner Brothers to the Right Honourable Earl of Kimberly," February 15, 1856, CO 147/19, The National Archives, Kew, UK.

10. Maps include *West Africa: Lagos (Now in Nigeria); Plan of the Marina Area, Showing Streets, Blocks of Buildings and the Owners of Property,* April 13, 1871, MR 1/389/4, 1 inch to 66 feet (scale), The National Archives, Kew, UK; J. Hunter (colonial engineer), *West Africa: Lagos (Now in Nigeria); Plan of Banner Bros & Co. Property,* August 7, 1871, 1 inch to 10 feet (scale), MR 1/389/2, The National Archives, Kew, UK.
11. "The Attorney General v. John Holt & Co., and Others and the Attorney General v. W. B. McIver & Co., and Others," in *Nigeria: Law Reports; A Selection from the Cases Decided in the Full Courts of the Colonies of Southern Nigeria and Nigeria,* vol. 2, *1911–1914* (Lagos: Government Printer, 1916), 1–70.
12. Robert Forsyth Irving, *A Collection of the Principal Enactments and Cases Relating to Titles to Land in Nigeria, Compiled in Connection with the Sale of Enemy Properties, Revised, with Introduction and Notes* (London: Stevens and Sons, 1916), xiii.
13. See W. T. G. Lawson (native of West Africa, assistant colonial surveyor), *Plan of the Town of Lagos, West Coast of Africa: Prepared for Lagos Executive Commissioners of the Colonial and Indian Exhibition 1886; By Order of Fred. Evans, Esq. CMG. Deputy Governor,* December 30, 1885, 1 inch to 275 feet (scale), CO 700/Lagos 9 (1), The National Archives, UK; *Plan of the Town of Lagos, West Africa (In 15 Sheets),* 1891, CO 700/Lagos 14, The National Archives, Kew, UK; *Plan of the Town of Lagos,* revised to May 1908, 1 inch to 5 chains (scale), CO 700/Lagos 31, The National Archives, Kew, UK.
14. Robert Forsyth Irving, "D. W. Lewis and Others v. Bankole and Others," in *Principal Enactments,* 217–38.
15. For the lagoons in the Bight of Benin, see Robin Law, "Between the Sea and the Lagoons: The Interaction of Maritime and Inland Navigation on the Precolonial Slave Coast," *Cahiers d'Études Africaines* 29 (1989): 209–37; Patrick Manning, "Coastal Society in the Republic of Bénin: Reproduction of a Regional System," *Cahiers d'Études Africaines* 29, no. 114 (1989): 239–57. Further west, see, for instance, Emmanuel Kwaku Akyeampong, *Between the Sea and the Lagoon: An Eco-social History of the Anlo of Southeastern Ghana, c. 1850 to Recent Times* (Athens: Ohio University Press; Oxford: James Currey, 2001).
16. Obaro Ikime, *The Fall of Nigeria: The British Conquest* (London: Heinemann, 1977); see also Alan Burns, *History of Nigeria* (London: Allen & Unwin, 1963); Sir William N. M. Geary, *Nigeria under British Rule* (New York: Barnes & Noble, 1965); and James Smoot Coleman, *Nigeria: Background to Nationalism* (Benin City: Broburg & Wiström, 1986).
17. J. F. Ade Ajayi, "The British Occupation of Lagos, 1851–1861: A Critical Review," *Nigeria Magazine* 69 (1961): 96–105.
18. Samuel Adjai Crowther, *Vocabulary of the Yoruba Language* (London: Printed for the Church Missionary Society, 1843), 284.
19. See Saburi O Biobaku, *The Egba and Their Neighbours 1842–1872* (Oxford: Clarendon, 1957).

20. For different perspectives through time, see, for instance, John M'Leod's description in 1803: "Along that line of coast from Cape St Paul's to Cape Formosa (an extent of, at least, three hundred miles) an easy communication between the various countries is afforded by means of lakes and rivers, which run in a direction nearly parallel with the sea-beach." M'Leod, *A Voyage to Africa with Some Account of the Manners and Customs of the Dahomian People* (London: Cass, 1971), 127. Almost fifty years later, in 1849, Commander Forbes wrote: "On the west side the lagoons may be said to join the Volta, although in the dry season, at a little distance from the town of Godomey (fifteen miles from Whydah), a sandy neck divides the Lagoons of Lagos and Whydah. Emptying into these lagoons are several navigable rivers, as yet but imperfectly known, except to slave enterprise." Frederick E. Forbes, *Dahomey and the Dahomans: Being the Journals of Two Missions to the King of Dahomey and Residence at His Capital, in the Years 1849 and 1850* (London: Cass, 1966), 1:9.
21. Very few works on the Bight of Benin use the lagoon as a centerpiece. For an unpublished example, see Jane Diane Matheson, "Lagoon Relations in the Era of Kosoko, 1845–1862: A Study of African Reaction to European Intervention" (PhD diss., Boston University, 1974). See also Caroline Sorenson-Gilmour, "Slave Trading along the Lagoons of South West Nigeria: The Case of Badagry," in *Ports of the Slave Trade (Bights of Benin and Biafra): Papers from a Conference of the Centre of Commonwealth Studies, University of Stirling, June 1998,* ed. Robin Law and Silke Strickrodt (Stirling, UK: University of Stirling, 1999), 84–95. For detailed descriptions of the coastal system, see Manning, "Coastal Society"; and Law, "Between the Sea and the Lagoons."
22. For instance, H. P. Ward's map, accurate orientation and soundings of the water depth, concentrated mostly on the "River Lagos." There is only an outline of Lagos Island with indications of the thick belt of mangrove trees that surround the island, and the only item in the otherwise blank section of the city is the king's palace on the northeastern tip of the island. This map, made in service of the "gunboat imperialism" that prompted the bombing of Lagos, inadvertently pinpoints the political and spatial focal point of the city, and its destruction was the signal to the indigenous people that they had lost, causing Kosọkọ and his people to flee the scene.
23. J. B. Harley, "Deconstructing the Map," in *Classics in Cartography: Reflections on Influential Articles from Cartographica,* ed. Martin Dodge (Chichester, UK: J. Wiley & Sons, 2011), 11, 12.
24. S. Max Edelson, *The New Map of Empire: How Britain Imagined America before Independence* (Cambridge, MA: Harvard University Press, 2017), 6.
25. See Barbara E. Mundy, *The Mapping of New Spain: Indigenous Cartography and the Maps of the Relaciones Geográficas* (Chicago: University of Chicago Press, 1996); Matthew H. Edney, *Mapping an Empire: The Geographical Construction of British India, 1765–1843* (Chicago: University of Chicago

Press, 1997); Ian J. Barrow, *Making History, Drawing Territory: British Mapping in India, c. 1756–1905* (New Delhi: Oxford University Press, 2003); Raymond B. Craib, *Cartographic Mexico: A History of State Fixations and Fugitive Landscapes* (Durham, NC: Duke University Press, 2004); Rebecca Solnit and Rebecca Snedeker, *Unfathomable City: A New Orleans Atlas* (Berkeley: University of California Press, 2013); Palmira Brummett, *Mapping the Ottomans: Sovereignty, Territory, and Identity in the Early Modern Mediterranean* (New York: Cambridge University Press, 2015); Edelson, *New Map of Empire;* and William Rankin, *After the Map: Cartography, Navigation, and the Transformation of Territory in the Twentieth Century* (Chicago: University of Chicago Press, 2018).

26. Mark S. Monmonier, *How to Lie with Maps* (Chicago: University of Chicago Press, 1991).
27. This map can be seen in books on Lagos, including Jean Herskovits Kopytoff, *A Preface to Modern Nigeria: The "Sierra Leonians" in Yoruba, 1830–1890* (Madison: University of Wisconsin Press, 1965); Kunle Akinsemoyin and Alan Vaughan-Richards, *Building Lagos,* 2nd ed. (Saint Helier, Jersey: Pengrail, 1977); John Godwin and Gillian Hopwood, *Sandbank City: Lagos at 150* (Lagos: Prestige, 2012); and Olasupo Shasore, *Possessed: A History of Law and Justice in the Crown Colony of Lagos 1861–1906* (Nigeria: Quramo Books, 2014).
28. "The Colonial and Indian Exhibition: West Africa," *Times,* July 17, 1886.
29. Karin Barber, *The Anthropology of Texts, Persons and Publics: Oral and Written Culture in Africa and Beyond* (Cambridge: Cambridge University Press, 2007), 1.
30. These categories are discussed in detail in "Dots and Lines on a Map: A Note on Method," which follows this chapter.
31. Other important documentary evidence includes the official correspondence generated by the agents of British empire, for instance, the many men engaged by the Preventative Squadron, a fleet patrolling the West African coast in a bid to stem the transatlantic trade. This book also draws on a variety of photographs, drawings, and paintings as historical evidence. There are also a host of transcribed indigenous sources, as well as the journals and testimonies of missionaries active in the region who were working for organizations such as the British Church Missionary Society.
32. Orhan Pamuk, *Istanbul: Memories of a City,* trans. Maureen Freeley (London: Faber and Faber, 2005).
33. *Sketch of Lagos and the Adjacent Country (to Accompany Report by Col. H. St. George Ord, R.E., 1865),* 1865, CO 700/Lagos 1, The National Archives, Kew, UK.
34. Ananya Roy, "Why India Cannot Plan Its Cities: Informality, Insurgence and the Idiom of Urbanization," *Planning Theory* 8, no. 1 (February 1, 2009): 76–87.
35. See chapter 1.

36. Lieut. J. H. Glover, R.N., *Lagos River,* 1859, 1:14,800 scale, FO 925/467, The National Archives, Kew, UK.
37. See the "Dots and Lines on a Map: A Note on Method" for the different categories of maps I use in the book.
38. Ademide Adelusi-Adeluyi, "Mapping Old Lagos: Digital Histories and Maps about the Past," *The Historian* 82, no. 1 (2020): 51–65.
39. Bamgboṣe Street is sometimes spelled "Gbamgboṣe" in historical texts and maps.
40. John Whitford, *Trading Life in Western and Central Africa* (Liverpool: The "Porcupine" Office, 1877), 88.
41. Abbé Borghero, *Lagos et Ses Environs: Faisant Suite à La Carte Du Royaume de Porto-Novo (Dans Le Vicariat Apostolique Des Côtes de Bénin)* 1871, about 2.5 inches to 1 mile (scale), CO 700/WestAfrica 20/2, The National Archives, Kew, UK.
42. Herbert Samuel Heclas Macaulay, *Justitia Fiat: The Moral Obligation of the British Government to the House of King Docemo of Lagos; An Open Letter* (London: Printed by St. Clements Press, 1921), 26.
43. Kristin Mann took notes on these grants in 1985, and this set is the only complete version of Ọba Dosunmu's grants available. I have been able to corroborate the full text of six of them from the records in the Foreign Office and Colonial Office files, in various land reports, and in Martin Robinson Delany's *Official Report of the Niger Valley Exploring Party* (New York: Webb, Millington; London: Thomas Hamilton, 1861).

DOTS AND LINES ON A MAP

1. Vincent Brown, "Mapping a Slave Revolt: Visualizing Spatial History through the Archives of Slavery," *Social Text* 125 (December 2015): 134–41.
2. Raymond B. Craib, *Cartographic Mexico: A History of State Fixations and Fugitive Landscapes* (Durham, NC: Duke University Press, 2004).
3. "Fundamentals of Georeferencing a Raster Dataset," ESRI, accessed March 7, 2019, http://desktop.arcgis.com/en/arcmap/10.3/manage-data/raster-and-images/fundamentals-for-georeferencing-a-raster-dataset.htm.
4. *Plan of the Town of Lagos—West Africa (in 15 Sheets) (Survey in Fourteen Sheets Numbered 1 to 13 and "11 & 15" and Bound as a Volume),* 1891, 1 inch to 88 feet (scale), CO 700/Lagos 14, The National Archives, Kew, UK; William C. Speeding, *Lagos Harbour,* April 1898, 1 inch to 1600 feet (scale), 63 × 90 cm, The National Archives, Kew, UK.
5. Speeding, *Lagos Harbour.*
6. "Fundamentals of Georeferencing."
7. Adam Jones, *German Sources for West African History, 1599–1669,* Studien Zur Kulturkunde (Wiesbaden: Franz Steiner Verlag GmbH, 1983), 23–24.
8. Lieutenant J. H. Glover, *Sketch of Lagos River,* 1859 (corrected to 1889) 1:20 scale, RGS MR Nigeria S.127–24, The Royal Geographical Society, London. This map also exists in a slightly different form as *Lagos River,* 1859,

Admiralty Chart 2812, 1:14,800 scale, FO 925/467, The National Archives, Kew, UK.

9. Sketch of Abeokuta, 1879, in Rev. Valentine Faulkner, *Information Respecting the Settlement of Lagos* (London: Colonial Office, 1879), CO 879/15/ African No. 192, The National Archives, Kew, UK.
10. Thomas Earl, J. D. Curtis, and W. H. Harris, *Africa West Coast: Lagos River,* 1857, Maps SEC.11.(2551.), The British Library; Thomas Earl, W. H. Harris, and J. D. Curtis, *The Entrance of the River Lagos,* November 1851, MR Nigeria S.127, The Royal Geographical Society, London.
11. Robert Norris, "Dahomy and Its Environs," 1789, in *Memoirs of the Reign of Bossa Ahádee, King of Dahomy, an Inland Country of Guiney, to Which Are Added the Author's Journey to Abomey, the Capital, and a Short Account of the African Slave Trade,* Cass Library of African Studies: Travels and Narratives, No. 32 (London: Cass, 1968).
12. *Africa, W Coast: Bight of Benin; Benin and Nigeria: Porto Novo to River Niger Delta; Item 4: "Point at Entrance of River Lagos from HMS Cyrene, Percy Grace Esq., Captain," Showing "Sketch A" and "B,"* ca. 1822, ADM 344/963/4, The National Archives, Kew, UK.
13. H. P. Ward, *A Plan of the River Lagos,* 1851, 3.5 inches to 1 sea mile (scale), FO 925/1052, The National Archives, Kew, UK.
14. *Plan of Lagos and the Disputed Lands Ordered by Consul Campbell: Inclosure in Slave Trade No. 31,* 1855, FO 84/976, The National Archives, Kew, UK; *Sketch Map of Akitoye's Grant to the CMS (with a Sketch of the Compounds),* 1855, FO 84/976, The National Archives, Kew, UK.
15. J. Hunter, *West Africa: Lagos (Now in Nigeria); Plan of the Marina Area, Showing Streets, Blocks of Buildings and the Owners of Property,* April 13, 1871, 1 inch to 66 feet (scale), MR 1/389/4, The National Archives, Kew, UK.
16. *Gold Coast Colony: Sketch Plan of the Town and Island of Lagos 1883; Compiled from Surveys by Sir John H. Glover and W. T. G. Lawson,* 1883, CO 700/Lagos 5, The National Archives, Kew, UK.
17. W. T. G. Lawson (native of West Africa, assistant colonial surveyor), *Plan of the Town of Lagos, West Coast of Africa: Prepared for Lagos Executive Commissioners of the Colonial and Indian Exhibition 1886; By Order of Fred. Evans, Esq. CMG. Deputy Governor,* December 30, 1885, 1 inch to 275 feet (scale), CO 700/Lagos 9 (1), The National Archives, UK; *Plan of the Town of Lagos,* 1891.
18. *Plan of the Town of Lagos,* revised to May 1908, CO 700/Lagos 31, The National Archives, Kew, UK.
19. *Plan of Lagos,* 1942, Maps X.1726, The British Library.
20. Lawson, *Plan of the Town of Lagos,* 1885; *Sketch of Lagos and the Adjacent Country (to Accompany Report by Col. H. St. George Ord, R.E., 1865),* 1865, CO 700/Lagos 1, The National Archives, Kew, UK.
21. J. H. Glover, *Africa—West Coast: Bight of Benin; Inland Water Communication between Laogs, Badagry, Porto Novo and Epè,* 1858–62, Maps SEC.11. (445a.), The Royal Geographical Society, London.

22. Alex Tickell, "Negotiating the Landscape: Travel, Transaction, and the Mapping of Colonial India," *Yearbook of English Studies* 34 (2004): 20.
23. Commander R. N. Bedingfield and Captain Richard F. Burton, *Africa—West Coast: River Ogun or Abbeokúta; Sketch Survey; Map (of Territory Now in South-West Nigeria) Showing the Course of the Ogun River,* March 24, 1862, 1 inch to about 3.14 miles (scale), March 24, 1862, MR 1/1804/4, The National Archives, Kew, UK; and Commander R. N. Bedingfield and Captain Richard F. Burton, *The Ogun (Abeokuta) River,* 1862, 1 square to 1 mile (scale), MR Nigeria S.141, The Royal Geographical Society, London; *Chart of the Ogun River, Nigeria, Showing Towns and Villages along Banks from Abeokuta to Lagos,* August 1892, MPG 1/850/4, The National Archives, Kew, UK.
24. Charles A. Gollmer to Henry Venn and Major Straith, January 15, 1853, CA2/O43, Church Missionary Society Archives: Yoruba Mission, Center for Research Libraries, crl.edu.
25. "Registry of Dosunmu's Land Grants, 1853–1861," Lagos Land Registry, based on notes taken by Kristin Mann, grant to James P. L. Davies from Ọba Dosunmu, November 13, 1860.
26. Consul Campbell to the Earl of Clarendon, Lagos, October 3, 1855, No. 13, in *Correspondence Relative to the Dispute between Consul Campbell and the Agents of the Church Missionary Society at Lagos* (printed for the use of the Foreign Office, 1856), 31, FO 403/5, National Archives, Kew UK.
27. S. Max Edelson, *The New Map of Empire: How Britain Imagined America before Independence* (Cambridge, MA: Harvard University Press, 2017), 10.
28. *Plan of the Town of Lagos,* 1891.
29. *Plan of the Town of Lagos,* May 1904; *Plan of the Town of Lagos,* August 1911, one inch to 5 chains (330 feet; scale), MR Nigeria S.13, The Royal Geographical Society, London; *Town of Lagos: West Africa (15 Sheets Mounted as One),* 1926, one inch to 88 feet (scale), CO 1047/676, The National Archives, Kew, UK.
30. *Plan of the Town of Lagos,* May 1904; *Plan of the Town of Lagos,* August 1911; *Town of Lagos: West Africa (15 Sheets Mounted as One).*
31. *Plan of Lagos,* 1942, Maps X.1726, The British Library.
32. Isale Eko Descendants' Union, *Isale Eko Day* (Isale Eko Descendants' Union, 2018), 32–33.
33. See *Plan of the Town of Lagos (Revised to August 1911),* 1911, MR Nigeria S.13, The Royal Geographical Society, London.

CHAPTER 1: STREETS, PLACEMAKING, AND HISTORY IN OLD LAGOS

1. Pẹlu Awofẹsọ is the author of numerous travel books, including *White Lagos* (Lagos: Homestead Media, 2017); *Nigerian Festivals: The Famous and Not So Famous* (Lagos: Homestead Media, 2013); *Tour of Duty: Journeys around Nigeria and Sketches of Everyday Life* (Lagos: Homestead Media, 2010); *A Place Called Peace: A Visitor's Guide to Jos, Plateau State, Nigeria* (Lagos: Homestead Travel Publications, 2003).

2. Visit https://newmapsoldlagos.com to see the maps.
3. For an overview of Glover's career in Lagos, see W. D. McIntyre, "Commander Glover and the Colony of Lagos, 1861–73," *Journal of African History* 4, no. 1 (March 1963): 57–79.
4. John Augustus Otonba Payne, *Table of Principal Events in Yoruba History, with Certain Other Matters of General Interest, Compiled Principally for Use in the Courts Within the British Colony of Lagos, West Africa* (Lagos: Printed by Andrew M. Thomas, 1893), 12.
5. James Baldwin, "Unnameable Objects and Unspeakable Crimes," in *The White Problem in America: First Published as a Special Issue of "Ebony Magazine," August, 1965*, ed. the Editors of Ebony (Chicago: Johnson Publishing, 1966), 174.
6. Tajudeen Sowole, "Furore as 190-Year-Old Monument Is Demolished," *The Guardian*, (Nigeria), September 29, 2016.
7. See chapter 2.
8. See, for instance, *Gold Coast Colony: Sketch Plan of the Town and Island of Lagos 1883; Compiled from Surveys by Sir John H. Glover and W. T. G. Lawson*, 1883, CO 700/Lagos 5, The National Archives, Kew, UK.
9. Adeoye Deniga, *Notes on Lagos Streets*, 3rd ed. (Lagos: Jehovah Shalom / Tika-Tore Printing Press, 1921), 3.
10. Samuel Crowther, ed., *A Vocabulary of the Yoruba Language: Compiled by the Rev. Samuel Crowther, Native Missionary of the Church Missionary Society; Together with Introductory Remarks by the Rev. O. E. Vidal, M. A. Bishop Designate of Sierra Leone* (London: Seeleys, 1852), 79.
11. In his map, the names are misspelled as "Abouta Aru," "Abouta Offee," and "Abouta Eggà." See Lieut. J. H. Glover, R.N., *Lagos River*, 1851, Royal Geographic Society, London.
12. See *Plan of the Town of Lagos—West Africa (in 15 Sheets)*, 1891, 1 inch to 88 feet (scale), CO 700/Lagos 14, The National Archives, Kew, UK.
13. Administrator Patey served in 1866; Administrator J. D. Dumaresq served in 1875.
14. See W. T. G. Lawson (Native of West Africa, Assistant Colonial Surveyor), *Plan of the Town of Lagos, West Coast of Africa: Prepared for Lagos Executive Commissioners of the Colonial and Indian Exhibition 1886; By Order of Fred. Evans, Esq. CMG. Deputy Governor*, 1885, CO 700/Lagos 9/1, The National Archives, Kew, UK; *Plan of the Town of Lagos*, 1891.
15. For visual evidence of these streets, see the fix-and-fill maps from 1926 and 1942: *"Town of Lagos. West Africa": 15 Sheets Mounted as One. Printed, Cloth. 88 Feet to an Inch. Cadastral Branch, Nigeria Surveys*, CO 1047/676, The National Archives, Kew, UK; *Plan of Lagos*, 1942, BL Maps X.1726, The British Library.
16. Olukoju adds further that chieftaincy titles like Aṣọgbọn and Bajulaiye trace their linguistic roots to Benin. See Ayọdeji Olukoju, "Which Lagos, Whose (Hi)Story?," *Lagos Notes and Records* 24, no. 1 (2018): 148.

17. See Ọba Dosunmu to the Earl of Kimberly, August 16, 1873, CO 147/28, The National Archives, Kew, UK.
18. Rev. J. Buckley Wood, "On the Inhabitants of Lagos: Their Character, Pursuits, and Languages," *Church Missionary Intelligencer and Record: A Monthly Journal of Missionary Information* 6, new series (November 1881): 683–91; Ọmọ-Ọba John B. Lọṣi, *Itan Eko* (Lagos: Church Missionary Society Bookshop, 1934); Sketch of Abeokuta, 1879, in Rev. Valentine Faulkner, *Information Respecting the Settlement of Lagos* (London: Colonial Office, 1879), CO 879/15 African No. 192, The National Archives, Kew, UK.
19. N. C. Mitchel, "Yoruba Towns," in *Essays on African Population,* ed. K. M. Barbour and R. Mansell Prothero (London: Routledge and Paul, 1961), 286.
20. Allan B. Jacobs, *Great Streets* (Cambridge, MA: MIT Press, 1993), 3.
21. Lieutenant J. H. Glover, *Sketch of Lagos River,* 1859 [corrected to 1889], 1:20 scale, MR Nigeria S.127–24, The Royal Geographical Society, London; William C. Speeding, F.G.R.S., Harbour Master, *Lagos Harbour,* April 1898, 1 inch to 1,600 feet (scale), CO 700/Lagos 19, The National Archives, Kew, UK.
22. Ologun Dudu, "A Lagos Account of Administrator Glover," *African Times: Journal of the African-Aid Society* 88, no. 88 (September 9, 1868): 44–45.
23. Deniga, *Notes on Lagos Streets,* 6.
24. Deniga, 7.
25. See *Africa West Coast: Lagos River,* surveyed by Mr. Earl, Master Lt. J. D. Curtis and Mr. William Harris, Acting Master, 1851, Maps SEC.11.(2551.), The British Library.
26. This location is currently Ahmadu Bello Way. In the nineteenth century, the bridge was opposite the old powder magazine on Eko. See *Gold Coast Colony: Sketch Plan of the Town and Island of Lagos;* Speeding, *Lagos Harbour;* and Kunle Akinsemoyin and Alan Vaughan-Richards, *Building Lagos,* 2nd ed. (Sain Helier, Jersey: Pengrail, 1977), 31.
27. Speeding, *Lagos Harbour,* April 1898.
28. *Map of Lagos and Environs,* 1932, National Museum Library and Archive.
29. *Certified Copies of the Acts of Lagos: 1862–1874,* 1862, CO 148/1, The National Archives, Kew, UK.
30. Jessica Marie Johnson, *Wicked Flesh: Black Women, Intimacy, and Freedom in the Atlantic World* (Philadelphia: University of Pennsylvania Press, 2020).
31. Lynda Nead, *Victorian Babylon: People, Streets and Images in Nineteenth-Century London* (New Haven, CT: Yale University Press, 2000), 16.
32. Liora Bigon, *A History of Urban Planning in Two West African Colonial Capitals: Residential Segregation in British Lagos and French Dakar (1850–1930)* (London: Edwin Mellen, 2009).
33. Rev. Charles A. Gollmer, "Original Communications: New Prospects of Usefulness at Lagos; Extract from Rev. Gollmer's Journal Dated Badagry, March 12, 1852," *Church Missionary Intelligencer: A Monthly Journal of Missionary Information* 3 (June 1852): 132–33.

34. The first publicly available map of Lagos, offered for sale in 1886, portrays a fully *streeted* city (at least in the western half of the island), presenting a stark contrast to Glover's 1859 map, which acknowledged only the indigenous quarters and the factories of European merchants. He saw "no streets to speak of" in his sketch, with the European factories seemingly floating in severely unarticulated space. The map is W. T. G. Lawson (Native of West Africa, Assistant Colonial Surveyor), *Plan of the Town of Lagos, West Coast of Africa: Prepared for Lagos Executive Commissioners of the Colonial and Indian Exhibition 1886; By Order of Fred. Evans, Esq. CMG. Deputy Governor,* 1885, CO 700/Lagos 9/1, The National Archives, Kew, UK. Ọba Dosunmu's land grants in the 1850s (examined in chapter 4) established a grammar for the city and helped ground the understanding of the initial expansion of quarters outside of the original six—Iga/Isalẹ Eko, Oko Faji, Olowogbowo, Ereko, Itọlọ, and Ọfin—into sites like Ẹhin Igbẹti, Ebute Alakoro, and Racecourse.
35. Dudu, "Lagos Account," 44–45.
36. Otonba John Augustus Payne, *Payne's Lagos and West African Almanack and Diary* (London: J. S. Phillips, 1882), 35.
37. Liora Bigon, "Urban Planning, Colonial Doctrines and Street Naming in French Dakar and British Lagos, c. 1850–1930," *Urban History* 36, no. 3 (2009): 426–48.
38. This title comes from Payne's "Names of Streets" reproduced in his almanacks. See, for instance, Payne, *Payne's Lagos and West African Almanack and Diary,* 1882, 35.
39. See, for instance, Payne, 35.
40. "Lagos: No. 14 (1863) Enclosure 2 in No. 14," in *The Reports Exhibiting the Past and Present State of Her Majesty's Colonial Possessions: Transmitted with the Blue Books for the Year 1863* (London: George Edward Eyre and William Spottiswoode for Her Majesty's Stationery Office, 1864), 45.
41. Ọba Dosunmu to the Earl of Kimberly, August 16, 1873.
42. Robert Campbell, *A Pilgrimage to My Motherland: An Account of a Journey among the Egbas and Yorubas of Central Africa, in 1859–60* (New York: Thomas Hamilton; Philadelphia: The Author, 1861), 43.
43. "Lagos: No. 14 (1863)," 41.
44. Richard F. Burton, "A Day at Lagos," in *Wanderings in West Africa from Liverpool to Fernando Po: By a F. R. G. S. with Map and Illustration* (London: Tinsley Brothers, 1863), 2:186–241.
45. Mrs. Henry Grant Foote, *Recollections of Central America and the West Coast of Africa: By Mrs. Foote, Widow of the Late Henry Grant Foote* (London: T. Cautley Newby, 1869), 190.
46. "Lagos: No. 14 (1863)."
47. "Administrator John Glover to Captain P. Sheppard, Administrator in Chief, Sierra Leone," August 5, 1871, CO 147/21, The National Archives, Kew, UK.

48. "Administrator John Glover to Captain P. Sheppard."
49. "Editorial," *The Anglo-African*, September 5, 1863.
50. A timeline of sorts can be constructed based on land grants issued from 1853 to 1861, then their records of conversion in the early 1860s in *The Anglo-African* newspaper.
51. "Lagos: No. 14 (1863) Enclosure 2 in No. 14," 45.
52. "Lagos: No. 14 (1863) Enclosure 2 in No. 14," 29.
53. "Lagos: No. 14 (1863) Enclosure 2 in No. 14," 26.
54. "Lagos: No. 9 (1867)," in *The Reports Showing the Present State of Her Majesty's Colonial Possessions: Transmitted with the Blue Books for the Year 1867; Part II. North American Colonies; African Settlements and St. Helena, &c.* (London: George Edward Eyre and William Spottiswoode for Her Majesty's Stationery Office, 1869), 26.
55. "Lagos: No. 11 (1866)," in *The Reports Exhibiting the Past and Present State of Her Majesty's Colonial Possessions: Transmitted with the Blue Books for the Year 1866, Part I. West Indies and Mauritius* (London: George Edward Eyre and William Spottiswoode for Her Majesty's Stationery Office, 1868), 29.
56. "Lagos: No. 4 (1871)," in *The Reports Showing the Present State of Her Majesty's Colonial Possessions: Transmitted with the Blue Books for the Year 1871* (London: George Edward Eyre and William Spottiswoode for Her Majesty's Stationery Office, 1869), 40.
57. Vox Populi [pseud.], "The Lagos Canal—One of the Great Holes in the Lagos Public Chest: To the Editor of the African Times," *African Times: Journal of the African-Aid Society* (January 6, 1872): 99.
58. "Lagos: No. 9 (1867)," 24.
59. "Lagos: No. 9 (1867)," 24.
60. "Lagos: No. 9 (1867)," 24.
61. "Lagos: No. 16 (1872)," in *Papers Relating to Her Majesty's Colonial Possessions, with Blue Book for 1872* (London: George Edward Eyre and William Spottiswoode for Her Majesty's Stationery Office, 1874), 136–40.
62. Osbert Chadwick, quoted in Thomas S. Gale, "Lagos: The History of British Colonial Neglect of Traditional African Cities," *African Urban Studies* 5 (Fall 1979): 15.
63. "Lagos: No. 14 (1863) Enclosure 2 in No. 14," 40.
64. Robert Campbell, "Editorial," *The Anglo-African*, September 19, 1863.
65. Campbell.
66. "Banner Brothers to the Right Honourable Earl of Kimberly," November 30, 1870, CO 147/19, The National Archives, Kew, UK.
67. See Burton and Bedingfield's itinerary maps for examples of the Ogun River. Comdr. R. N. Bedingfield and Capt. R. F. Burton, *The Ogun (Abeokuta) River*, 1862, MR Nigeria S.141, Royal Geographical Society, London.
68. "Copy of Original Title Deed of Banner Brothers and Co's Property at Lagos: Inclosure in Banner Brothers to the Right Honourable Earl of Kimberly," February 15, 1856, CO 147/19, The National Archives, Kew, UK.

69. "Title Deed of Banner Brothers."
70. See chapter 4.
71. "Banner Brothers to Earl of Kimberly," November 30, 1870.
72. "Banner Brothers to Earl of Kimberly," November 30, 1870.
73. J. Hunter (colonial engineer), *West Africa: Lagos (Now in Nigeria); Plan of Banner Bros & Co. Property,* MR 1/389/2, The National Archives, UK.
74. "Administrator John Glover to Captain P. Sheppard."
75. Rev. C. A. Gollmer to Rev. Henry Venn (Badagry), March 12, 1852, C/A2/O43, Church Missionary Society Archives: Yoruba Mission.
76. There is some confusion in the historiography about James White's origins, although he wrote in a journal that he was born in Sierra Leone. See his entry on December 31, 1854, in the journals of Rev. James White (African) in Badagry, Lagos, and Otta, 1850–79, C/A2/O87, Church Missionary Society Archive: Yoruba Mission. Olatunji Ojo writes that White was born in Owu and "enslaved in the 1820s" after which he was rescued and settled in Freetown. See Olatunji Ojo, "The Yoruba Church Missionary Society Slavery Conference 1880," *African Economic History* 49, no. 1 (2021): 98n5. Kopytoff writes that it was his parents who were enslaved in Owu and ended up liberated in Freetown, where he was born. See Jean Herskovits Kopytoff, *A Preface to Modern Nigeria: The "Sierra Leonians" in Yoruba, 1830–1890* (Madison: University of Wisconsin Press, 1965), 300.
77. Journals of Rev. James White, June 14, 1852.
78. Journals of Rev. James White, June 14, 1852.
79. Journals of Rev. James White, June 14, 1852.
80. Journals of Rev. James White, March 3–6, 1852.
81. *Sketch of Abeokuta,* CO 879/15 African No. 192; Rev. J. B. Wood, *Historical Notices of Lagos, West Africa and on the Inhabitants of Lagos: Their Character, Pursuits, and Languages* (Lagos: Church Missionary Society Bookshop, 1933), 57.
82. Otonba John Augustus Payne, *Payne's Lagos and West African Almanack and Diary* (London: J. S. Phillips, 1883), 21, 35.
83. Scala Square begins to appear on maps in 1885. In some maps, it is spelled "Scale" Square, meaning that it might have been named after a different merchant.
84. Deniga, *Notes on Lagos Streets,* 2.
85. *Correspondence Relative to the Dispute between Consul Campbell and the Agents of the Church Missionary Society at Lagos* (printed for the use of the Foreign Office, 1856), FO 403/5, The National Archives, Kew, UK.
86. Continuing, he wrote, "Application by the King (Akintoye) on behalf of the merchants she treats with contempt, setting his authority at defiance; yet this woman is a protege of Mr. Gollmer because she is an Egba woman." Quoted in Oladipo Yemitan, *Madame Tinubu: Merchant and King-Maker* (Ibadan: University Press, 1987), 18.

87. Quoted in Yemitan, *Madame Tinubu,* 20.
88. See fix-and-fill maps from 1885, 1891 and 1908:
89. She was said to reside in the place owned later by Alli Balogun. See Deniga, *Notes on Lagos Streets,* 5.
90. Advertisements for whole and broken bricks began appearing in *The Anglo-African* around 1864—for instance, on the front page of *The Anglo-African,* November 26, 1864.
91. For a full investigation of the effects of incendiarism in Lagos, see C. Onyeka Nwanunobi, "Incendiarism and Other Fires in Nineteenth-Century Lagos (1863–88)," *Africa: Journal of the International African Institute* 60 (1990): 111–20.
92. Otonba Payne recorded dozens of significant fires on the island in the nineteenth century. See Payne, *Table of Principal Events,* 8.
93. See Payne, 8.
94. *Certified Copies of the Acts of Lagos.*
95. This translates as "A fire that brought the heavens down."
96. Payne, *Table of Principal Events,* 8.
97. Payne, 8.
98. "Great Fires of Lagos" and "Lagos (from Our Own Correspondents)," *The African Times: Journal of the African-Aid Society* 11, no. 106 (February 23, 1863): 91.
99. "Editorial," *The Anglo-African,* October 28, 1865.
100. "Editorial."
101. See Nigeria Law Reports, "Inasa and Others v. Chief Sakariyawo Oshodi," in *A Selection of Cases Decided in Divisional Courts of Nigeria, and on Appeal in the Full Court and in the Privy Council during the Period 1932–1934,* vol. XI (Lagos: Government Printer, 1935), 9.
102. In one case, Amodu Inasa, one descendant, had been evicted in 1925 for his role in trying to sell land in Ẹpẹtẹdo. Nigeria Law Reports, "Inasa and Others v. Chief Sakariyawo Oshodi," 4. Additionally, around 1928, the government acquired land in the Ẹpẹtẹdo quarter, causing a rift between the families that had settled there. Nigeria Law Reports, "Inasa and Others v. Chief Sakariyawo Oshodi," 13.

    Some descendants of Ọṣodi Tapa sued, claiming that the compensation for the land should come to them since they were the original owners of the entire quarter. The respondents, Moriamo Dakolo and others, claimed instead that even though Ọṣodi family members were the original settlers, they had yielded the land to Dakolo's ancestors, who subsequently confirmed their ownership through a government grant.
103. Nigeria Law Reports, "Sakariyawo Oshodi v. Moriamo Dakolo and Others," in *A Selection of Cases Decided in Divisional Courts of Nigeria during the Years 1928 and 1929, Together with the Privy Council and Full Courts Appeals Relating to Cases Falling within These Two Years,* vol. 9 (Lagos: Government Printer, 1932), 14.

104. Nigeria Law Reports, "Sakariyawo Oshodi v. Moriamo Dakolo and Others," 14.
105. Nigeria Law Reports, "Sakariyawo Oshodi v. Moriamo Dakolo and Others," 14.
106. Nigeria Law Reports, "Sakariyawo Oshodi v. Moriamo Dakolo and Others," 16–17.
107. Nigeria Law Reports, "Inasa and Others v. Chief Sakariyawo Oshodi," 11.

CHAPTER 2: WHO BROKE LAGOS?

Epigraph: Ben Okri, *The Famished Road* (New York: Anchor Books, 1993), 3.

1. "Captain Jones to Commodore Bruce, 'Bloodhound,' off the North Point of Lagos: December 29, 1851 (Inclosure No. 2 in No. 70)," in *Papers Relative to the Reduction of Lagos by Her Majesty's Forces on the West Coast of Africa,* vol. XXXVII.533 [1455] (London: Harrison and Son, 1852), 197, PP 1852, LIV.221, Great Britain, Parliamentary Papers (House of Commons Sessional Papers).
2. W. T. G. Lawson, *Plan of the Town of Lagos, West Coast of Africa: Prepared for Lagos Executive Commissioners of the Colonial and Indian Exhibition 1886; By Order of Fred. Evans, Esq. CMG. Deputy Governor,* 1885, CO 700/Lagos 9/1, The National Archives, Kew, UK; and *Plan of the Town of Lagos—West Africa (in 15 Sheets),* 1891, 1 inch to 88 feet (scale), CO 700/Lagos 14, The National Archives, Kew, UK.
3. Thomas Earl, J. D. Curtis, and William Harris, *Africa West Coast: Lagos River,* 1851, Maps SEC.11 (2551), The British Library, London.
4. Earl, Curtis, and Harris, *Lagos River,* 1851.
5. "The State of the Global City," Urban History Association Plenary, October 2021.
6. Laurent Fourchard, "Between World History and State Formation: New Perspectives on Africa's Cities," *Journal of African History* 52, no. 2 (2011): 223–48.
7. These verses in English are "I dispersed one towards Popo land / And another towards Epe / one who overflowed [the] river / with the slaves of his father." John B. Prince Losi, *History of Lagos* (1914; repr., Lagos: African Education Press, 1967), 27.
8. For instance, in 1841, Kosọkọ was exiled after being defeated by Oluwọle. He zigzagged through the lagoons, ending up in Ouidah, when Akitoye called him back in 1842.
9. See Patrick Cole, *Modern and Traditional Elites in the Politics of Lagos* (Cambridge: Cambridge University Press, 1975).
10. According to Prince Kọtun's data, Kumaifo "capped" Kutere, Adele, and Oṣinlokun; Jose capped Oluwọle; Oṣobule capped Akitoye; and Jiyabi capped Kosọkọ. See Prince Bọlakalẹ Kọtun, *History of the Eko Dynasty* (Lagos: Allentown Limited, 2008), 186.
11. See Losi, *History of Lagos;* Patrick D. Cole, "Lagos Society in the Nineteenth Century," in *Lagos: The Development of an African City,* ed. A. B. Aderibigbe (Ikeja: Longman Nigeria, 1975), 27–58; Rev. John B. Wood, *Historical Notices of Lagos, West Africa and on the Inhabitants of Lagos: Their*

Character, Pursuits, and Languages (1878; repr., Lagos: Church Missionary Society Bookshop, 1933).

12. Kristin Mann, "African and European Initiatives in the Transformation of Land Tenure in Colonial Lagos (West Africa), 1840–1920," in *Native Claims: Indigenous Law against Empire, 1500–1920* (Oxford: Oxford University Press, 2012), 223–47.
13. See Cole, *Modern and Traditional Elites.*
14. See Robert Sydney Smith, *The Lagos Consulate, 1851–1861* (London: Macmillan; Lagos: University of Lagos Press, 1978); and Cole, *Modern and Traditional Elites.*
15. See Losi, *History of Lagos.*
16. There were also rumors that he had died by suicide, though it was never substantiated.
17. Kọtun identified five contenders, all of whom were descendants of the Ologun Kutere: Akitoye, Olukoya, Akiolu, Kosọkọ, and Olosi. See Kọtun, *History of the Eko Dynasty,* 42.
18. J. F. Ade Ajayi, "The British Occupation of Lagos, 1851–1861: A Critical Review," *Nigeria Magazine: The Lagos Centenary Issue* 69 (1961): 99.
19. Ọmọ-Ọba John B. Lọṣi, *Itan Eko* (Lagos: Church Missionary Society Bookshop, 1934), 31. Published in English as Prince Losi, *History of Lagos.*
20. Kọtun, *History of the Eko Dynasty,* 45.
21. Lieutenant J. W. Glover, *Sketch of Lagos River,* 1859 (corrected to 1889) 1:20 scale, RGS MR Nigeria S.127–24, The Royal Geographical Society, London.
22. See David A. Ross, "The Career of Domingo Martinez in the Bight of Benin 1833–64," *Journal of African History* 6 (1965): 79–90.
23. Robin Law, "The Career of Adele at Lagos and Badagry, c. 1807–c. 1837," *Journal of the Historical Society of Nigeria* 9, no. 2 (1978): 35–59.
24. Rev. Samuel Annear, quoted in Jane Diane Matheson, "Lagoon Relations in the Era of Kosoko, 1845–1862: A Study of African Reaction to European Intervention" (PhD diss., Boston University, 1974).
25. Kosọkọ was Oṣinlokun's son. Oṣinlokun had ruled Lagos through the 1820s. "Kings of Lagos during the Last Fifty Years: Compiled by Rev. Charles Gollmer of the Church Missionary Society; Lagos, September 2, 1853," FO 84/920, The National Archives, Kew, UK.
26. See Rev. Charles A. Gollmer, "Western Africa, Extracts from the Journal of Rev. Charles Gollmer," *Missionary Register* 34, no. 6 (June 1846).
27. Sketch of Abeokuta, 1879, in Rev. Valentine Faulkner, *Information Respecting the Settlement of Lagos* (London: Colonial Office, 1879), CO 879/15 African No. 192, The National Archives, Kew, UK.
28. "Though every hour brings some new information respecting the ill-intents of our enemies towards us." Extract from the journal of S. Annear, dated October 1845, *Wesleyan Missionary Notices,* August 1846.
29. Extract from the journal of S. Annear, dated August 1845, *Wesleyan Missionary Notices,* August 1846.

30. Extract from the journal of S. Annear, dated August 1845, *Wesleyan Missionary Notices,* August 1846.
31. Extract from the journal of S. Annear, dated August 1845, *Wesleyan Missionary Notices,* August 1846.
32. See Rev. Charles A. Gollmer, "Rebellion at Lagos—Deposition of the King," *Missionary Register* 34 (1846): 436–46.
33. Rev. Charles A. Gollmer wrote, "Our premises are at the extreme east end of Badagry, on the Lagoon side." Gollmer, "Western Africa," 267.
34. Rev. Henry Townsend, "Western Africa, Extracts from the Journal of Rev. Henry Townsend, Dated September 7, 1845," *Missionary Register* 34 (1846).
35. Losi, *History of Lagos,* 12
36. Gollmer, "Rebellion at Lagos," 439.
37. Losi, *History of Lagos,* 12
38. Robert Forsyth Irving, *A Collection of the Principal Enactments and Cases Relating to Titles to Land in Nigeria, Compiled in Connection with the Sale of Enemy Properties, Revised, with Introduction and Notes* (London: Stevens and Sons, 1916), 52.
39. "Akitoye, the King, left Lagos the night after Eletu fled, and went to Abẹokuta, where he at present remains. He is an intimate friend of Wawu, the English Chief there. The Portuguese suffered the loss of nearly all their property at Lagos. The whole town is nearly destroyed, and the numerous inhabitants reduced to but a few." Gollmer, "Rebellion at Lagos."
40. "We have heard indirectly that they [Abẹokuta and his men] are busily employed in building a wall around the town, with a moat on the outside. Perhaps they have been induced to do this, on account of the report which circulated a week or two since, that the Dahoman [*sic*] power intended to direct its strength against their metropolis, and, by conquering it, subdue the whole country to their sway." Rev. Samuel Annear, "Missions in Guinea and the Slave Coast: Extract of a Letter from Rev. Samuel Annear, Dated Badagry May 18, 1845," *Wesleyan-Methodist Magazine,* July 1845.
41. See Annear.
42. See Saburi O. Biobaku, *The Egba and Their Neighbours, 1842–1872* (Oxford: Clarendon Press, 1957), 33.
43. Rev. Samuel Annear, "The Wesleyan Mission at Badagry," *Wesleyan-Missionary Notices* 9, no. 105 (September 1847).
44. Rev. Henry Townsend, "Return of Messrs. Marsh and Williams from Abbekuta—Visit of the Missionaries to Akitoye, at Imowo—His Entry to Badagry—First Visit of a Chief to the Sea," *Missionary Register* 34, no. 10 (1846): 441.
45. Gollmer, "Rebellion at Lagos."
46. Gollmer, 442.
47. "Obba Shoron to Captain Jones: Badagry, July 3, 1851 (Inclosure 18 in No. 41)," in *Papers Relative to the Reduction of Lagos,* 130. Jones replied: "You are aware that the object of England is, 'peace and goodwill to all countries,'

and that we do not interfere with the internal arrangements of other people, but leave them to settle their own affairs, so long as British subjects are protected and unmolested by the Government of the country in which they reside." "Captain Jones to Obba Shoron: 'Sampson' July 18, 1851 (Inclosure 19 in No. 41)," in *Papers Relative to the Reduction of Lagos,* 131.

48. Harmattan winds in the Bight of Benin bring cool, foggy mornings before the intense heat of the midday sun. One visitor noted in the 1850s that the "Harmattan winds prevail about Christmas time. They are very dry and cold: I have seen at 8 A.M., the thermometer at 54° Fahr., during the prevalence of these winds. The mornings and evenings, however warm the noon might be, are always comfortable. The general range of the temperature is between 74° and 90° Fahr. I have experienced warmer days in New York and Philadelphia." See Robert Campbell, *A Pilgrimage to My Motherland: An Account of a Journey among the Egbas and Yorubas of Central Africa, in 1859–60* (New York: Thomas Hamilton; Philadelphia: The Author, 1861), 138.
49. "Commodore Bruce to the Secretary of the Admiralty, '*Penelope,*' Off Lagos, January 2, 1852," in *Papers Relative to the Reduction of Lagos,* 194.
50. Kosọkọ and his men had defeated the first wave of attacks by British ships in November 1851.
51. "The Rev. C. A. Gollmer to Commodore Fanshawe: Badagry, March 26, 1851 (Inclosure 4 in No. 36)," in *Papers Relative to the Reduction of Lagos,* 105.
52. Ann Laura Stoler, ed., *Imperial Debris: On Ruins and Ruination* (Durham, NC: Duke University Press, 2013), 9.
53. See "Commodore Bruce to the Secretary of the Admiralty, '*Penelope,*' Ascension, January 17, 1852," in *Papers Relative to the Reduction of Lagos.*
54. "Captain Jones to Commodore Bruce," 197.
55. "Consul Beecroft to Viscount Palmerston, 'Bloodhound,' off Lagos, January 3, 1852 (No. 69)," in *Papers Relative to the Reduction of Lagos,* 187.
56. After the bombardment, these ship names and those of some of their captains were attached to the city. Wilmot Point, in honor of the captain of the *Harlequin,* is still in use today at the southwestern edge of Victorian Island. These names are visible in Ward's sketch of the river in 1851: *A Plan of the River Lagos by Mr H. P. Ward, Second Master, HMS Sampson,* 1851, 3.5 inches to 1 sea mile (scale), FO 925/1052, The National Archives, Kew UK.
57. A contemporary term describing the water bodies surrounding the island.
58. Earlier treaties had been signed peacefully, and by force after the successful bombardment of Gallinas.
59. "King Akitoye to Consul Beecroft (circa February 1851)," in *Papers Relative to the Reduction of Lagos.*
60. Rev. Jonathan Layton Buckley Wood of the Church Missionary Society was a missionary stationed in Lagos and Abẹokuta from 1860 to 1880. (See his papers in C/A2/O96, The National Archives, Kew, UK.) He wrote the first documented precolonial history of Lagos, spanning from the original

settlement to its annexation by the British government on August 6, 1861. The Colonial Office first published it in June 1879.

61. See in full: *Papers Relative to the Reduction of Lagos.*
62. "The Destruction of Lagos, on the Coast of Africa, by the British Squadron," *Illustrated London News,* February 21, 1852.
63. "Captain Jones to Commodore Bruce," 196, 197.
64. "Captain Jones to Commodore Bruce: 'Sampson,' off Lagos, January 1, 1852," in *Papers Relative to the Reduction of Lagos,* 206.
65. The next chapter follows this process of repopulating and rebuilding.
66. "Consul Beecroft to Viscount Palmerston," 189.
67. Adeoye Deniga, *Notes on Lagos Streets,* 3rd ed. (Lagos: Jehovah Shalom / Tika-Tore Printing, 1921), 4.
68. Benjamin N. Lawrance, Emily Lynn Osborn, and Richard L. Roberts, "Introduction: African Intermediaries and the 'Bargain' of Collaboration," in *Intermediaries, Interpreters, and Clerks: African Employees in the Making of Colonial Africa,* Africa and the Diaspora (Madison: University of Wisconsin Press, 2006), 3–34.
69. Frederick Edwyn Forbes, *Dahomey and the Dahomans: Being the Journals of Two Missions to the King of Dahomey, and Residence at His Capital* (London: Longman, Brown, Green, and Longmans, 1851), 2:176–77.
70. "Captain Jones to Commodore Bruce," 195.
71. "Captain Jones to Commodore Bruce," 197.
72. "Commander Wilmot to Commodore Bruce: 'Harlequin' off Lagos, December 1, 1851 (Inclosure 2 in No. 76)," in *Papers Relative to the Reduction of Lagos,* 211.
73. See map of Kuramo Island in William C. Speeding, F.G.R.S., Harbour Master, *Lagos Harbour,* April 1898, 1 inch to 1,600 feet (scale), CO 700/Lagos 19, The National Archives, Kew, UK.
74. Mark S. Monmonier, *Drawing the Line: Tales of Maps and Cartocontroversy* (New York: H. Holt, 1995), 57.
75. "Commodore Wilmot to Commander T. G. Forbes: 'Harlequin,' Lagos, November 26, 1851 (Inclosure in No. 65)," in *Papers Relative to the Reduction of Lagos,* 171.
76. Rev. Charles A. Gollmer, "Original Communications: New Prospects of Usefulness at Lagos; Extract from Rev. Gollmer's Journal Dated Badagry, March 12, 1852," *Church Missionary Intelligencer: A Monthly Journal of Missionary Information* 3 (June 1852): 133.
77. Quote from section heading: *Plan of the Town of Lagos, West Africa (in 15 Sheets),* 1891, 1 inch to 88 feet (scale), Sheets 3–8, CO 700/Lagos 14, The National Archives, Kew, UK; "Captain Jones to Commodore Bruce," 197.
78. *A Plan of the River Lagos by Mr H. P. Ward;* also "A Plan of the River Lagos (Inclosure 14 in No. 70)," in *Papers Relative to the Reduction of Lagos,* facing p. 204.
79. "Consul Beecroft to Commodore Bruce, 'Bloodhound,' off Lagos, November 27, 1851: Inclosure 2 in No. 67," in *Papers Relative to the Reduction of Lagos,* 184.

80. "Consul Beecroft to Commodore Bruce," 185.
81. See Robin Law, "Trade and Politics behind the Slave Coast: The Lagoon Traffic and the Rise of Lagos, 1500–1800," *Journal of African History* 24 (1983): 321–48; Kristin Mann, *Slavery and the Birth of an African City: Lagos, 1760–1900* (Bloomington: Indiana University Press, 2007); Suzanne Miers and Igor Kopytoff, *Slavery in Africa: Historical and Anthropological Perspectives* (Madison: University of Wisconsin Press, 1977).
82. See chapter 1.
83. Irving, *Collection of the Principal Enactments,* 288.
84. Irving, 288.
85. "The evidence shows that the place was unoccupied until the reign of Akitoye, but between 1841 and 1845 the bush was cleared by one of the war chiefs of Kosoko's party, named Oshodi, who occupied the land with his domestics until the English bombardment at the end of 1851, when they were driven out. In course of time the dwellers in that neighbourhood began to protect their property by driving in stakes, locally known as 'cabbage posts,' and filling up to them." Robert Forsyth Irving, "The Peace Concluded between Dosumu and Kosọkọ: Letter from Dr. Irving, Jan. 30, 1854, Detailing the Event," *Church Missionary Intelligencer: A Monthly Journal of Missionary Information,* May 1854, 288.
86. Bishop Street has been renamed Issa Williams Street now. See maps .google.com.
87. Deniga, *Notes on Lagos Streets,* 3.
88. Also mentioned on the morning of December 27 in "Consul Beecroft to Viscount Palmerston," 188.
89. "Consul Beecroft to Viscount Palmerston," 187.
90. *A Plan of the River Lagos by Mr H. P. Ward;* also "A Plan of the River Lagos (Inclosure 14 in No. 70)."
91. See "Registry of Dosunmu's Land Grants, 1853–1861, Lagos Land Registry," based on notes taken by Kristin Mann.
92. Robert Forsyth Irving, "D. W. Lewis and Others v. Bankole and Others," in *Collection of the Principal Enactments,* 296.
93. See Akin L. Mabogunje, "Lagos: The Rise of a Modern Metropolis," in *Urbanization in Nigeria* (New York: Africana Pub., 1969), 238–73; Liora Bigon, *A History of Urban Planning in Two West African Colonial Capitals: Residential Segregation in British Lagos and French Dakar (1850–1930)* (London: Edwin Mellen, 2009).
94. "Captain Jones to Commodore Bruce," 197.
95. "Commodore Bruce to the Secretary of the Admiralty, 'Penelope,' Off Lagos, January 2, 1852," 194.
96. This story comes from Nina Mba's chapter "Women in Lagos Political History," in *History of the Peoples of Lagos State,* ed. Ade Adefuye, Babatunde Agiri, and Jide Osuntokun (Lagos: Lantern Books, 2002), 233–45. Adding to her mythology, the Ọṣodi family contends that in keeping with tradition, Ọba Akinsemoyin asked Chief Aromire to give her land near an iroko tree, where

she could build a shrine, and thence an iga. Iga Faji, her palace, does not appear on any maps, but the specific site of the market begins to appear in the 1880s, for instance, in Lawson's map of the city. Interestingly enough, Glover's 1859 map notes one prominent tree close to the present site of the Iga Faji. But there is also an alternative narrative about Faji contained in the correspondence of the WALC (West African Land Commission), on page 247.

97. In fact, there were several reports of returnees being robbed and even murdered in Lagos.
98. "Rev. Henry Townsend to Captain H. D. Trotter, Communicated to Viscount Palmerston by Captain Trotter, April 7, 1851, Abẹokuta, December 10, 1850 (No. 3)," in *Papers Relative to the Reduction of Lagos*, 88.
99. "King Akitoye to Consul Beecroft (circa February 1851)," 97–98.
100. "Consul Beecroft to Viscount Palmerston," 187.
101. "Minute of a Conference with Kosoko, Chief of Lagos (Inclosure 2 in No. 55)," in *Papers Relative to the Reduction of Lagos*, 148.
102. James George Philp, *British Men o' War Attacked by the King of Lagos*, 1851, oil on canvas, Government Art Collection, https://artcollection.culture.gov.uk/artwork/6427/.
103. "Captain Jones to Commodore Bruce."
104. Quoted in Charles A. Gollmer, "Lagos, and Missionary Operations in the Bight of Benin," *Church Missionary Intelligencer and Record: A Monthly Journal of Missionary Information* 4, no. 12 (December 1853): 268.
105. Frazer to Clarendon, S.T. No. 9, undated, FO 84/920, The National Archives, Kew, UK.
106. "King Akitoye to Consul Beecroft (circa February 1851)."
107. Robert Smith points to how Campbell adopted a more conciliatory stance toward Kosọkọ. See Smith, *Lagos Consulate*, 58.
108. "Kosoko to Campbell," written in Portuguese by Tapa, English translation provided by Campbell, September 7, 1853, FO 84/920.
109. For information on the management of the blockades, stalling of palm oil exports and the issue of kidnapping on the lagoons, see Smith, *Lagos Consulate*, 56–58.
110. Irving, "Peace Concluded between Dosumu and Kosọkọ," 106–9.
111. Irving, 106–9.

CHAPTER 3: A NEW EKO?

Epigraph: "Vice Consul Louis Fraser to the Earl of Malmesbury: Slave Trade No. 5," March 11, 1853, FO 84/920, The National Archives, UK.

1. Rev. Martin to Vice Consul Fraser, Lagos, December 27, 1852, Slave Trade No. 5, FO 84/920, The National Archives, Kew, UK.
2. The first Government House, or "iron coffin," was built in 1853 next to the CMS house at the intersection of Water Street and Ọdunlami Street. Finally, a larger structure was built south of the racecourse on the Marina between Cable Street and Force Road. See maps.google.com.

3. See James White, "Original Communications: New Prospects of Usefulness at Lagos; Extract from James White's Journal Dated January 10, 1852," *Church Missionary Intelligencer: A Monthly Journal of Missionary Information* 3 (June 1852): 124–25; and Kristin Mann, *Slavery and the Birth of an African City: Lagos, 1760–1900* (Bloomington: Indiana University Press, 2007), 93.
4. *Plan of Lagos and the Disputed Lands Claimed by the Church Missionary Society,* inclosure in Slave Trade No. 31, 1855, FO 84/976, The National Archives, Kew, UK.
5. See *Plan of Lagos and the Disputed Lands.*
6. Evidence in Coelho v. Pereira, Chief Magistrate's Court, 1864, Supreme Court Records, Lagos, quoted in Antony G. Hopkins, "Property Rights and Empire Building: Britain's Annexation of Lagos, 1861," *Journal of Economic History* 40, no. 4 (December 1980): 786.
7. Olatunji Ojo, "Document 2: Letters Found in the House of Kosoko, King of Lagos (1851)," *African Economic History* 40, no. 1 (2012): 37–126.
8. "Copy of Grant of Land by King Akitoye to the Church Missionary Society," in *Correspondence Relative to the Dispute between Consul Campbell and the Agents of the Church Missionary Society at Lagos* (London: Printed for the use of the Foreign Office, 1856), inclosure 7 in No. 6, pp. 11–12, FO 403/5, The National Archives, Kew, UK.
9. Mr. Sandeman to Consul Campbell, Lagos, August 28, 1855 (Inclosure 1 in No 11), in *Correspondence Relative to the Dispute between Consul Campbell and the Agents of the Church Missionary Society at Lagos* (London: Printed for the use of the Foreign Office, 1856), 23, FO 403/5, The National Archives, Kew, UK.
10. Rev. J. Buckley Wood, "On the Inhabitants of Lagos: Their Character, Pursuits, and Languages," *Church Missionary Intelligencer and Record: A Monthly Journal of Missionary Information* 6 (November 1881): 683–91; Paul Osifodunrin, *The First Church in Lagos: History of Holy Trinity Church, Ebute Ero, 1852–2016* (Ibadan: University Press PLC, 2018), 16.
11. "Copy of Grant of Land by King Akitoye to the Church Missionary Society (Inclosure 7 in No 6)," in *Correspondence Relative to the Dispute,* 12.
12. "'Death of Akitoye' in Western Africa—Church Missionary Society—Yoruba Mission," *Missionary Register,* June 1854, 277.
13. Prince Bọlakalẹ Kọtun, *A History of the Eko Dynasty* (Abuja: National Assembly Printing Press, 1998), 187.
14. See "Funeral Obsequies of King Akitoye," in "Western Africa—Church Missionary Society: Lagos," *Missionary Register,* June 1855, 265.
15. "Funeral Obsequies of King Akitoye," 265. He used the term "idolator" here, which likely included the Muslim population as well.
16. "Atchooboo" and "Kosae" are the chiefs who signed the treaty with him, but they remain unidentified.
17. "Treaty with the King and Chiefs of Lagos, Signed January 1, 1852," in *Lagos: Additional Papers Relating to the Occupation of Lagos; Presented to*

the House of Lords by Command of Her Majesty, 1862 (London: Harrison & Sons, 1862), 1:1–2, PP LXI.339, 365, Great Britain, Parliamentary Papers (House of Commons Sessional Papers).

18. Smith talks about the twelve treaties signed on the West African coast in this period.
19. "Engagement with the King and Chiefs of Lagos (Inclosure in No. 69)," in *Papers Relative to the Reduction of Lagos by Her Majesty's Forces on the West Coast of Africa*, vol. XXXVII.533 [1455] (London: Harrison and Son, 1852), 190–92, PP 1852, LIV.221, Great Britain, Parliamentary Papers (House of Commons Sessional Papers).
20. The ninth article provided accommodation for France to sign on to the treaty.
21. Article II in "Treaty with the King and Chiefs of Lagos, Signed January 1, 1852," 1.
22. Article V in "Treaty with the King and Chiefs of Lagos, Signed January 1, 1852," 1.
23. Article VIII in "Treaty with the King and Chiefs of Lagos, Signed January 1, 1852," 2.
24. Article VIII in "Treaty with the King and Chiefs of Lagos, Signed January 1, 1852," 2.
25. "Engagement with the King and Chiefs of Lagos," 190–92.
26. Article VIII in "Treaty with the King and Chiefs of Lagos, Signed January 1, 1852," 2.
27. Article VI in "Treaty with the King and Chiefs of Lagos, Signed January 1, 1852," 2.
28. Though most of the merchants were British, other nationalities, such as Austrian, Portuguese and German, were also represented in the treaty.
29. Articles I and II in "Agreement with the King and Chiefs of Lagos, Signed February 28, 1852: Engagement between King Akitoye and His Chiefs, and British and Other European Merchants Trading at Lagos," in *Lagos: Additional Papers Relating to the Occupation of Lagos; Presented to the House of Lords by Command of Her Majesty, 1862* (London: Harrison and Son, 1862), 2, PP LXI.339, 365, Great Britain, Parliamentary Papers (House of Commons Sessional Papers).
30. Article III in "Agreement with the King and Chiefs of Lagos, Signed February 28, 1852," 3.
31. Article III in "Agreement with the King and Chiefs of Lagos, Signed February 28, 1852," 3.
32. Article X in "Agreement with the King and Chiefs of Lagos, Signed February 28, 1852," 3.
33. Article II (following consideration) in "Agreement with the King and Chiefs of Lagos, Signed February 28, 1852," 4.
34. They were employed as interpreter and clerk, respectively.
35. Consul Campbell to the Earl of Clarendon, Lagos, April 4, 1855 (No 1), in *Correspondence Relative to the Dispute,* 1.

36. Consul Campbell to the Earl of Clarendon, Lagos, April 4, 1855 (No 1), 1.
37. Charles Gollmer to Vice Consul Fraser, February 28, 1853, FO 84/920, The National Archives, Kew, UK.
38. Consul Campbell to the Earl of Clarendon, Lagos, May 28, 1855, No. 6, in *Correspondence Relative to the Dispute*, 5–12.
39. Louis Fraser to Earl of Malmesbury, Lagos, March 11, 1853, FO 84/920, The National Archives, Kew, UK.
40. Louis Fraser to Earl of Malmesbury, Lagos, March 11, 1853.
41. Consul Campbell to the Earl of Clarendon, Lagos, September 1, 1855 (No. 11), in *Correspondence Relative to the Dispute*, 20–21.
42. See "Registry of Dosunmu's Land Grants, 1853–1861," Lagos Land Registry, based on notes taken by Kristin Mann, grant to Clara Maria Lisboa from Ọba Dosunmu, June 7, 1855.
43. Consul Campbell to the Earl of Clarendon, October 22, 1855, FO 84/976, The National Archives, Kew, UK. 1855.
44. See Consul Campbell to the Earl of Clarendon, Lagos, October 25, 1855, FO 84/976.
45. Consul Campbell to the Earl of Clarendon, Lagos, October 25, 1855.
46. Consul Campbell to the Earl of Clarendon, Lagos, October 25, 1855.
47. Rev. Martin to Vice Consul Fraser.

CHAPTER 4: RECOVERING LOST GROUND IN OLD LAGOS

Epigraph: Herbert Samuel Heclas Macaulay, "Speech Delivered in the Glover Memorial Hall, Lagos, on Saturday the 10th of August, 1940, at a General Meeting of the Nigerian National General Meeting of the Nigerian National Democratic Party, by the Founder, Mr. Herbert Macaulay, C.E., on the Burning Question of the Public Cemeteries on the Island of Lagos" (n.d.), file 4, box 17—Cemeteries Question, Herbert Macaulay Papers, Special Collections, University of Ibadan.

1. See "Registry of Dosunmu's Land Grants, 1853–1861," Lagos Land Registry, based on notes taken by Kristin Mann.
2. *The Anglo-African*, September 19, 1863. Administrative records show that in 1868, 717 "natives" and four Europeans died. In 1869, the number was 671 to 10. In 1870, it was 703 to 2. In 1871, it was 620 to 9. And in 1872 the number of Lagosians who died dropped to 531, while the number of Europeans stayed at 9. Herbert Samuel Heclas Macaulay, "Speech Delivered in the Glover Memorial Hall, Lagos, on Saturday the 10th of August, 1940, at a General Meeting of the Nigerian National Democratic Party, by the Founder, Mr. Herbert Macaulay, C.E., on the Burning Question of the Public Cemeteries on the Island of Lagos" (n.d.), p. 5, Herbert Macaulay Papers, File 4, Box 17—Cemeteries Question, University of Ibadan, Special Collections.
3. See "Registry of Dosunmu's Land Grants, 1853–1861."
4. "Reports from the Chief Magistrate's Court: Civil Cases; Ferreira vs. Cardosa, November 6, 1865," *The Anglo-African*, November 11, 1865.

5. "Reports from the Chief Magistrate's Court."
6. "Reports from the Chief Magistrate's Court."
7. Kristin Mann, *Slavery and the Birth of an African City: Lagos, 1760–1900* (Bloomington: Indiana University Press, 2007), 239.
8. How the other classes of chiefs obtained their land is less historically clear. See "No. 69. Report: Land Tenure in the Colony of Lagos. Appendix I: Statement of the Prince Eleko," in *African (West) No. 1048: West African Lands Committee; Committee on the Tenure of Land in the West African Colonies and Protectorates; Correspondence and Papers Laid before the Committee* (London: Colonial Office, 1916), 238.
9. See T. C. Rayner and John J. C. Healy, *Land Tenure in West Africa: Reports by T. C. Rayner, Esq., Chief Justice of Lagos, and J. J. C. Healy, Esq., Land Commissioner* (London: Foreign and Commonwealth Office Collection, 1897); Sir Meryn L. Tew, *Report to Title to Land in Lagos: Together with a Report of a Committee Set Up to Advise the Governor in Regard Thereto, and Draft Legislation to Give Effect to Certain Recommendations Contained Therein* (Lagos: Government Printer, 1939).
10. I am grateful to Dr. Kristin Mann for so generously sharing this and other documents with me.
11. Herbert Samuel Heclas Macaulay, *The Lagos Land Question: Speech Delivered on the Occasion by Mr. Herbert Macaulay, at Government House, Lagos, 13th June, 1912* (Lagos, 1912).
12. Macaulay, 14.
13. "Registry of Dosunmu's Land Grants, 1853–1861."
14. The text for grants to Diedrichserr and Grotto, James Cole, and Muribeca can be found in John J. C. Healy, "Land Tenure in the Colony of Lagos," in *Land Tenure in West Africa*, 5–6. Delany's grant can be seen in Delany, *Official Report of the Niger Valley Exploring Party* (New York: Webb, Millington; London: Thomas Hamilton, 1861), 29. Thomas C. Cole's grant is enclosed in Freeman to Duke, October 29, 1862, CO 147/1, The National Archives, Kew, UK.
15. See grants to de Souza and to J. G. Bartos in "Registry of Dosunmu's Land Grants, 1853–1861."
16. See grants to R. A. Johnson in "Registry of Dosunmu's Land Grants, 1853–1861."
17. John Augustus Otonba Payne, *Table of Principal Events in Yoruba History, with Certain Other Matters of General Interest, Compiled Principally for Use in the Courts within the British Colony of Lagos, West Africa* (Lagos: Printed by Andrew M. Thomas, 1893), 17.
18. Robert Forsyth Irving, "D. W. Lewis and Others v. Bankole and Others," in *A Collection of the Principal Enactments and Cases Relating to Titles to Land in Nigeria, Compiled in Connection with the Sale of Enemy Properties, Revised, with Introduction and Notes* (London: Stevens and Sons, 1916), 217.
19. Irving, 232.

20. See chapter 2 for the discussion of the bombardment and its immediate consequences in Lagos.
21. Irving, "D. W. Lewis and Others v. Bankole and Others," 233.
22. See Akin L. Mabogunje, "Lagos, a Study in Urban Geography" (PhD diss., University of London, 1961).
23. "Registry of Dosunmu's Land Grants, 1853–1861."
24. One exception is the August 1855 grant to Groto et al., in "Registry of Dosunmu's Land Grants, 1853–1861."
25. See "Registry of Dosunmu's Land Grants, 1853–1861."
26. There is little evidence of the origin of the Portuguese quarter, but it appears in maps in the 1870s and 1880s, such as W. T. G. Lawson, *Plan of the Town of Lagos, West Coast of Africa: Prepared for Lagos Executive Commissioners of the Colonial and Indian Exhibition 1886; By Order of Fred. Evans, Esq. CMG. Deputy Governor,* CO 700/Lagos 9/1. The National Archives, Kew, UK. Its name is perhaps a reference to the Portuguese slave traders who lived in Lagos during Kosọkọ's reign, or more likely a reference to the Brazilians settling there and Portuguese emerging as the dominant language in the quarter.
27. "Registry of Dosunmu's Land Grants, 1853–1861."
28. See chapter 3 for more on James Sandeman.
29. See chapter 3 for the circumstances around the Church Missionary Society's land grants after the bombardment.
30. Robert Sydney Smith, *The Lagos Consulate: 1851–1861* (London: Macmillan; Lagos: University of Lagos Press, 1978), xx.
31. Emigration Office to Elliot, November 11, 1862, CO 147/2, The National Archives, Kew, UK.
32. See Healy, "Land Tenure in the Colony of Lagos."
33. See Macaulay, *Lagos Land Question.*
34. Delany, *Official Report of the Niger Valley Exploring Party,* 29.
35. Macaulay, *Lagos Land Question,* 10.
36. Macaulay, 10.
37. Macaulay, 10.
38. Healy, "Land Tenure in the Colony of Lagos," 3.
39. The remaining three grants that either do not fit these templates or have not been identified are McCoskry's grant to George Roberts on December 24, 1861 and two unidentified grants.
40. See Healy, "Land Tenure in the Colony of Lagos," 1–14.
41. Healy, 5.
42. "Treaty with the King and Chiefs of Lagos, Signed January 1, 1852," in *Lagos: Additional Papers Relating to the Occupation of Lagos; Presented to the House of Lords by Command of Her Majesty, 1862* (London: Harrison & Sons, 1862), 1:1–2, PP LXI.339, 365, Great Britain, Parliamentary Papers (House of Commons Sessional Papers).
43. See William C. Speeding, *Lagos Harbour,* April 1898, 1 inch to 1,600 feet (scale), CO 700/Lagos 19, The National Archives, Kew, UK.

44. See "Registry of Dosunmu's Land Grants, 1853–1861."
45. Rev. Samuel Ajayi Crowther, *A Vocabulary of the Yoruba Language* [. . .] (London: Seeleys, Fleet Street, 1852).
46. See, for instance, in Lawson, *Plan of the Town of Lagos.*
47. See "Registry of Dosunmu's Land Grants, 1853–1861."
48. Healy, "Land Tenure in the Colony of Lagos," 5.
49. "Consul Campbell to the Earl of Clarendon."
50. "Consul Campbell to the Earl of Clarendon."
51. See Mann's notes on the land grants in the Registry Book.
52. Alhaji Lawal Babatunde Adams, *"Counting the Planets": The History and Development of OkePopo Community* (Lagos: Akuro Enterprises, 2003), 1–2.
53. Adams, 2, 3.
54. Elevation information comes from this 1904 map of Lagos, which contains contour lines for the elevation of the island: *Plan of the Town of Lagos (1904, Revised to May 1908)*, 1904, CO 700/Lagos 31, The National Archives, Kew, UK. This map has the Oke Popo area ranging approximately seven to thirteen feet above sea level, in comparison to the Isalẹgangan Lagoon, which is approximately three feet above sea level.
55. "No. 69. Report: Land Tenure in the Colony of Lagos," 234. T. G. O. Gbadamosi has identified *Lemomu* as an Arabic loan word in Yoruba that means "imam." See the appendix in T. G. O. Gbadamosi, *The Growth of Islam among the Yoruba, 1841–1908*, Ibadan History Series (Atlantic Highlands, NJ: Humanities Press, 1978).
56. Mann, *Slavery and the Birth of an African City*, 251.
57. Emigration Office to Elliot.
58. Emigration Office to Elliot.
59. This ordinance is missing from the record of acts in Lagos usually found in "Certified Copies of the Acts of Lagos: 1862–1874," CO 148/1, The National Archives, Kew, UK, but is referred to in CO 147 and reprinted in Irving, *Collection of the Principal Enactments and Cases*, 1–9.
60. Mann, *Slavery and the Birth of an African City*, 252.
61. Mann, 252.
62. For instance, William I. J. Chapman, the Land Commissioner's Court clerk, put out an announcement dated November 2, 1863, in *The Anglo-African* stating that successful applicants were invited to pick up the grants from 10 a.m. to noon every day, "except Friday and Sundays." See "Notice," *The Anglo-African*, November 31, 1863.
63. Macaulay, *Lagos Land Question*, 11.
64. *The Anglo-African* was published weekly on Saturdays.
65. "Notice," *The Anglo-African*, November 31, 1863.
66. Macaulay, *Lagos Land Question*, 11.
67. Macaulay, 12.
68. A British Subject, "Communication: To the Editor of The Anglo-African," *The Anglo-African*, September 19, 1863.

Epigraph: Public Opinion [pseud.], "To the Editor of the African Times," *African Times: Journal of the African-Aid Society* 9, no. 101 (December 23, 1869): 64.

1. An early grant converted in 1865 was listed in *The Anglo-African* as being in "Tinaboo's Square."
2. *Plan of the Town of Lagos, West Africa (in 15 Sheets),* 1891, 1 inch to 88 feet (scale), CO 700/Lagos 14, The National Archives, Kew, UK; William C. Speeding, *Lagos Harbour,* April 1898, 1 inch to 1,600 feet (scale), CO 700/ Lagos 19, The National Archives, Kew, UK.
3. Gov. Stanhope Freeman to the Duke of Newcastle: Lagos No. 7, March 8, 1862, CO 147/1, The National Archives, Kew, UK. Freeman wrote: "With regards to the courts of law, I have temporarily established four: A police court, criminal, and slave courts, and a commercial tribunal." He added: "The Police Court is open every day and settles all petty cases, committing for trial by the Criminal Court persons against whom any more serious charges are brought."
4. Fred I. A. Omu, "The Anglo-African 1863–65," *Nigeria Magazine,* September 1966, 206–12.
5. *Freedom Park . . . the Journey of Liberation: A Commemorative Publication to Mark the Commissioning of Freedom Park, the Transformed Old Broad Street Prison as Part of Lagos State's Official Celebration of Nigeria at 50* (Lagos: Hermitage Publishing, 2010), inside cover.
6. Lagos: No. 9 (1869), 24, CO 150, The National Archives, Kew, UK.
7. CO 148/1, Lagos Acts, The National Archives, Kew, UK.
8. "This has been our cry from the very day that Lagos was annexed and became a British settlement. We said it was her mission to bring a reflection of European and Christian civilization home to the very borders of the Egba and Yoruba countries. Civilization is catching; and we now look forward with even more than former confidence." "Editorial: What Can Be Done in Africa?," *African Times: Journal of the African-Aid Society* 9, no. 102 (January 24, 1870): 79.
9. "Bridewell" is a reference to a notorious prison in England.
10. *Ologun Dudu* means "Black warrior" in English. He could also have been Willioughby, Glover's right-hand man, if he was indeed a Saro civil servant.
11. Quoted in Tekena N. Tamuno, *The Police in Modern Nigeria, 1861–1965: Origins, Development and Role* (Ibadan: Ibadan University Press, 1970), 12.
12. "Editorial: Outrage on Mr. Turner," *African Times: Journal of the African-Aid Society* 9, no. 93 (April 23, 1869): 119–20.
13. "Editorial: Shameful Mal-administration of Justice at Lagos—the Chief Magistrate Suspended," *African Times: Journal of the African-Aid Society* 9, no. 101 (December 23, 1869): 68–69.
14. "Gov. Stanhope Freeman to the Duke of Newcastle."

15. J. P. L. Davies notes a time when even the ọba's wives were nearly assaulted when some sailors landed opposite the palace and tried to rush into the women's quarters.
16. "Editorial: How to Conciliate Natives, Etc. in a New Colony," *African Times: Journal of the African-Aid Society* 3, no. 30 (December 23, 1863): 68.
17. "Editorial: How to Conciliate Natives," 68.
18. "Lagos (from Our Correspondents)," *African Times: Journal of the African-Aid Society* 2, no. 20 (February 23, 1863): 91–92.
19. One of the Defendants [pseud.], "To the Editor of the African Times," *African Times: Journal of the African-Aid Society* 3, no. 33 (March 23, 1864): 106–7.
20. Broad-Street Hatch [pseud.], "How Justice Is Done at Lagos: To the Editor of the African Times," *African Times: Journal of the African-Aid Society* 6, no. 83 (May 23, 1868): 133.
21. Spencer H. Brown, "Colonialism on the Cheap: A Tale of Two English Army Surgeons in Lagos, Samuel Rowe and Frank Simpson, 1862–1882," *International Journal of African Historical Studies* 27, no. 3 (1994): 551–88.
22. See "Illegal Sentences," *African Times: Journal of the African-Aid Society* 8, no. 85 (July 23, 1868): 2.
23. Lagos [pseud.], "European Example in West Africa—Brutal Criminal Assault on a Child," *African Times: Journal of the African-Aid Society* 2, no. 8 (April 23, 1868): 117. See also An African [pseud.], "Justice or No Justice," *African Times: Journal of the African-Aid Society* 3, no. 85 (July 23, 1868): 9.
24. Brown, "Colonialism on the Cheap."
25. Broad-Street Hatch, "How Justice Is Done at Lagos," 133.
26. Lagos, "European Example in West Africa," 117.
27. Testis [pseud.], "Gross Outrage by Administrator Glover: To the Editor of the African Times," *African Times: Journal of the African-Aid Society* 9, no. 101 (December 23, 1869): 116–17.
28. "Affidavit of Mr. Turner, One of the Victims: Enclosure in Gross Outrage by Administrator Glover; To the Editor of the African Times," *African Times: Journal of the African-Aid Society* 9, no. 101 (April 23, 1869): 117.
29. Testis, "Gross Outrage," 116–17.
30. "Affidavit of Mr. Turner, One of the Victims," 117.
31. "Editorial: Outrage on Mr. Turner," 119.
32. Testis, "Gross Outrage," 120.
33. "Editorial: What Can Be Done in Africa?," 78–79.
34. Public Opinion [pseud.], "To the Editor of the African Times," *African Times: Journal of the African-Aid Society* (November 16, 1869): 64.
35. Public Opinion, "To the Editor of the African Times," November 16, 1869, 64.
36. Consequence [pseud.], "Horrible Brutality on Board the Steamer Thomas Bazley," *African Times: Journal of the African-Aid Society* 9, no. 101 (December 23, 1869): 65.

37. Public Opinion [pseud.], "To the Editor of the African Times," December 23, 1869, 64.
38. William Reffle, "Dinner to Ex-Chief Magistrate Way," *African Times: Journal of the African-Aid Society* 9, no. 106 (May 23, 1870): 125.
39. Consequence, "Horrible Brutality," 64–65.
40. Consequence [pseud.], "The Failure of Justice at Lagos—Wreck of the Steamer Thomas Bazley," *African Times: Journal of the African-Aid Society* 9, no. 102 (January 24, 1870): 77.
41. Reffle, "Dinner to Ex-Chief Magistrate Way," 125.
42. Reffle, 125.
43. Consequence, "The Failure of Justice at Lagos," 77.
44. Consequence, "The Failure of Justice at Lagos," 77.
45. Public Opinion, "To the Editor of the African Times," December 23, 1869, 64.
46. W. D. McIntyre, "Commander Glover and the Colony of Lagos, 1861–73," *Journal of African History* 4, no. 1 (March 1963): 58.
47. McIntyre, 59n11.
48. McIntyre, 60.
49. Lady Elizabeth Rosetta Glover, *Life of Sir John Hawley Glover, R.N. G.C.M.G.* (London: Smith, Elder, 1897), 79.
50. Glover, 78.
51. McIntyre, "Commander Glover," 59, 61.
52. They also wrote: "Lagos needs improvement not tomtoming and showing our powers, by going about the rivers and creeks with our steamer and steam launch to frighten poor people out of their wits by night and day. Our Administrator should look after Lagos, and not be spending our money in making useless presents to the interior kings; otherwise, we shall soon be bankrupt." Watchman [pseud.], "Another Lagos Letter: To the Editor of the African Times," *African Times: Journal of the African Aid Society* 11, no. 126 (December 23, 1871): 64–65.
53. McIntyre, "Commander Glover."
54. "Ex King Kossokkoh to His Excellency, John H. Glover RN: Lagos, Inclosure 1 in No. 19," February 29, 1872, CO 147/23, The National Archives, Kew, UK.
55. "Ex King Kossokkoh to His Excellency."
56. "Aboriginal Natives of Lagos to the Right Honorable the Earl of Kimberly: Inclosure 2 in No. 19," February 28, 1872, CO 147/23, The National Archives, Kew, UK.
57. "Aboriginal Natives of Lagos to the Right Honorable the Earl of Kimberly."
58. Others included Odunsi Ọmọ Ọba (a prince), Olumole, Olorisha Ọba Dosunmu, Okolo Balogun Ọba Dosunmu, Odunlami, Kossokoh Yesufu, Shitta, Odunlami of Oṣodi Street, and Sogunro ti Dosunmu. Women included Faji, several Ainas, three Pelus, Onitolla, Mohorunkeji, Osematu, Ajahtu, Enikolokiki, Awobiyi Erelu, Dada, Fayemi, Kubolaku, etc.

59. Pope Hennessy to Earl of Kimberly, 1871, CO 147/24, The National Archives, Kew, UK.
60. Pope Hennessy to Earl of Kimberly.
61. Ologun Dudu [pseud.], "Good Governance at Lagos," *African Times: Journal of the African-Aid Society* 9, no. 101 (December 23, 1869): 64.
62. Africa [pseud.], "The Growing Evils of Lagos," *African Times: Journal of the African-Aid Society* 11, no. 126 (December 23, 1871): 69–70.
63. The Eko People [pseud.], "Inclosure in Letter to Earl Kimberly on the Reported Stoppage of the Abeokuta Roads: Copy of a Letter Dated Lagos, March 16," *African Times: Journal of the African-Aid Society* 11, no. 130 (April 23, 1872): 134.
64. The Eko People, 134.
65. Liars Are Being Found Out [Am-Bay-Kay-Mon] [pseud.], "Arrival of the Governor-in-Chief, and Departure of Governor Glover for England; To the Editor of the African Times," *African Times: Journal of the African-Aid Society* 11, no. 133 (July 23, 1872): 99.
66. Negro [pseud.], "Fears of Danger through Executive Errors at Lagos: To the Editor of the African Times," *African Times: Journal of the African-Aid Society* 11, no. 126 (December 23, 1871): 64.
67. Africa, "The Growing Evils of Lagos," 69–70.
68. Quoted in McIntyre, "Commander Glover," 75.
69. Otonba John Augustus Payne, *Payne's Lagos and West African Almanack and Diary* (London: J. S. Phillips, 1882).

CONCLUSION

Epigraph: Wole Soyinka, "Unsinkable City: Reflections on a Lifetime in Lagos," in *Stranger's Guide: Lagos, Nigeria; The Past and Future City* (2020), 7.

1. Catherine Coquery-Vidrovitch, "The Process of Urbanization in Africa (from the Origins to the Beginning of Independence)," *African Studies Review* 34, no. 1 (1991): 4.
2. Rashauna Johnson, *Slavery's Metropolis: Unfree Labor in New Orleans during the Age of Revolutions* (Cambridge, MA: Cambridge University Press, 2016), 4.
3. Guadalupe Garcia, *Beyond the Walled City: Colonial Exclusion in Havana* (Oakland: University of California Press, 2016), 11.
4. See, for instance, Abdul Malik Simone, *For the City Yet to Come: Changing African Life in Four Cities* (Durham, NC: Duke University Press, 2006); and Abdul Malik Simone, "People as Infrastructure," in *Johannesburg: The Elusive Metropolis,* ed. Achille Mbembe and Sarah Nuttall (Durham, NC: Duke University Press, 2008), 68–90.
5. See Achille Mbembe, *On the Postcolony* (Berkeley: University of California Press, 2001); Achille Mbembe and Sarah Nuttall, "Writing the World from an African Metropolis," *Public Culture Public Culture* 16, no. 3 (2004):

347–72; and Sarah Nuttall and Achille Mbembe, *Johannesburg: The Elusive Metropolis* (Durham, NC: Duke University Press, 2004).

6. He has done so most recently in his work on Ghana: Ato Quayson, *Oxford Street: City Life and the Itineraries of Transnationalism* (Durham, NC: Duke University Press, 2014).
7. David M. Anderson and Richard Rathbone, “Introduction: Urban Africa; Histories in the Making,” in *Africa's Urban Past,* ed. David M. Anderson and Richard Rathbone (Oxford: James Currey, 2000), 9.
8. Simi Olorunfemi, “Eko,” in *Lagos of the Poets,* ed. Odia Ofeimun (Lagos: Hornbill House of the Art, 2010), 242.
9. Examples come from the work of Legacy95, LovingLagos, and the Yellow of Lagos.

Bibliography

ARCHIVAL MATERIAL

Bodleian Library, Oxford University
    Lagos Maps.
The British Library, London
    Maps SEC.11.
    Maps Y.484.
    Maps X.1726.
Cambridge University Library, Special Collections
    Lagos Maps
    RCMS Royal and Commonwealth Royal Commonwealth Society Collection
Church Missionary Society Archive. Center for Research Libraries. https://www.crl.edu.
    Section IV: Africa Missions
        C/A2/O43. The Journals of Rev. Charles Andrew Gollmer, Badagry, Lagos, and Abeokuta, Including Table of the Kings of Lagos from about the Year 1800, with Historical Observations, Etc. 1845–62.
        C/A2/O87. The Journals of Rev. James White (African) in Badagry, Lagos, and Otta, 1850–79.
Great Britain, Parliamentary Papers (House of Commons Sessional Papers)
    1842.XI.1, XII.1, Select Committee on the West Coast of Africa, Report, Minutes of Evidence, Appendices, Index.
    1852.LIV.221, Papers Relative to the Reduction of Lagos, 1851.
    1858–9.XXXIV.281, Class B, Correspondence with Foreign Powers Relating to the Slave Trade (PPRL).
    1862.LXI.339, Papers Relating to the Occupation of Lagos, 1861.
    1862.LXI.339, 365. Additional Papers Relating to the Occupation of Lagos, 1861.
    1865.V.1, Report from the Select Committee Appointed to Consider the State of British Establishments on the Western Coast of Africa, Proceedings, Minutes of Evidence, Appendix, Index.

1865.XXXVII.287, Report of Col. Ord, Commissioner to Inquire into the Condition of British Settlements on the West Coast of Africa.
High Court, Lagos State, Lagos
"Registry of Dosunmu's Land Grants, 1853–1861." Lagos Land Registry. Based on notes preserved by Kristin Mann.
The National Archives, Kew, UK
Colonial Office
CO 147, Lagos, Original Correspondence.
CO 148, Lagos, Acts.
CO 700, Maps and Plans: Series I.
CO 879, War and Colonial Department and Colonial Office: Africa, Confidential Print.
CO 1047, Maps and Photographs: Series II.
CO 1069, Maps and Photographs.
MR: Maps and Plans
MR 1, Public Record Office: Maps and plans extracted to rolled storage from various series of records
Foreign Office
FO 2, Political and Other Departments: General Correspondence before 1906, Africa.
FO 84, Slave Trade Department and Successors: General Correspondence before 1906.
FO 403, Dispute between Consul Campbell and the Agents of the Church Missionary Society at Lagos. Correspondence.
National Museum, Lagos
Map Collections.
Special Collections.
The Nigerian National Archives, University of Ibadan
BADADIV 1–8 Badagry Divisional Office (1865–1937).
CSO 1, Colony of Lagos. Despatches to the Colonial Office (63 volumes: 1861–1906).
Herbert Macaulay Papers.
Provincial and Divisional Office Papers of the Former Colony and Protectorate of Lagos.
Published Primary Sources
Blue Books, Lagos Colony, 1862–1905.
The Royal Geographical Society, London
MR. Nigeria Series.

Maps

Ajele and Extension. In Adams and Alli, *Oko Faji,* 272.
Amido Konsult. *Map Lay Out of Ita Tinubu Community.* In Adams and Alli, *Oko Faji,* 273. Middletown, DE: CreateSpace Independent Publishing Platform, 2017.

Bedingfield, Commander R. N., and Captain Richard F. Burton. *Africa—West Coast: River Ogun or Abbeokúta; Sketch Survey. Map (of Territory Now in South-West Nigeria) Showing the Course of the Ogun River.* March 24, 1862. 1 inch to 3.14 miles (scale). MR 1/1804/4. The National Archives, Kew, UK.
———. *The Ogun (Abeokuta) River.* 1862. 1 square to 1 mile (scale). MR Nigeria S.141. The Royal Geographical Society, London.
Borghero, Abbé. *Carte de la Côte des Esclaves.* 1865. 1:920,000. FO 925/454. The National Archives, Kew, UK.
———. *Lagos et Ses Environs: Faisant Suite à La Carte Du Royaume de Porto-Novo (Dans Le Vicariat Apostolique Des Côtes de Bénin).* 1871. 2.5 inches to 1 mile (scale). CO 700/WestAfrica 20/2. The National Archives, Kew, UK.
Chart of the Ogun River, Nigeria, Showing Towns and Villages along Banks from Abeokuta to Lagos. August 1892. MPG 1/850/4. The National Archives, Kew, UK.
Cotton, E. P. *Map of Town and Island of Lagos.* 1904. 1 inch to 20 chains (scale). CO 700/Lagos 27. The National Archives, Kew, UK.
Earl, Thomas, W. H. Harris, and J. D. Curtis. *The Entrance of the River Lagos.* November 1851. MR Nigeria S.127. The Royal Geographical Society, London.
———. *Lagos River.* 1851. Maps SEC.11.(2551) 1:18,100. The British Library, London.
Glover, John H. *Bight of Benin: Inland Water Communication between Lagos, Badagry, Porto Novo and Epè.* 1858–59 and 1862. Maps SEC.11.(445a.). Royal Geographical Society, London.
———. *Lagos River.* 1859. Admiralty Chart 2812. 1:14,800 scale. FO 925/467. The National Archives, Kew, UK.
———. *Sketch of Lagos River.* 1859 [corrected to 1889]. 1:20 scale. MR Nigeria S.127–24. The Royal Geographical Society, London.
———. *Surveys of Lagos and the River Niger 1859–1862 (in 124 Sheets).* 1862. Nigeria-Spec. River Niger. S.39 Portfolio 66. The Royal Geographical Society, London.
Gold Coast Colony: Sketch Plan of the Town and Island of Lagos 1883: Compiled from Surveys by Sir John H. Glover and W. T. G. Lawson. 1883. CO 700/Lagos 5. The National Archives, Kew, UK.
Hazzan, Arch. *The Traditional Map of Okepopo Community.* In *The History, People and Culture of Ita-Tinubu Community,* 10. Lagos: Tinubu Foundation, 2002.
———. *Traditional Map of Isalegangan Community.* In Adams and Alli, *Oko Faji,* 273.
Hunter, J. *Plan of Banner Bros. & Co. and Showing Part of Badagry.* August 3, 1871. MS, colored, 1 inch to 40 feet, 70 × 46 cm. MR 1/389/3. The National Archives, Kew, UK.
———. *Plan of Banner Bros. & Co. Property.* July 22, 1871. 1 inch to 10 feet (scale), 138 × 74 cm. MR 1/389/2. The National Archives, Kew, UK.
———. *West Africa: Lagos (Now in Nigeria); Plan of the Marina Area, Showing Streets, Blocks of Buildings and the Owners of Property.* April 13, 1871. 1 inch to 66 feet (scale). MR 1/389/4. The National Archives, Kew, UK.

Ita Tinubu Community. In Adams and Alli, *Oko Faji,* 271.
Lagos. 1909. CO 1047/534. The National Archives, Kew, UK.
Lagos and Central Africa Railway First Section from Lagos to Abbeokuta: Map (of Territory Now in South-Western Nigeria) Showing the Route from Abeokuta, Approximately Parallel with the Ogun River, as Far South as the Northern Shore of Lagos Lagoon. 1880. 1 inch to 1 nautical mile (scale). MR 1/1804/2. The National Archives, Kew, UK.
Lawson, W. T. G. *Plan of Proposed Line of Embankment, Lagos (Now in Nigeria): Showing the Lagoon, Church Mission, Government House, and Other Buildings, with Their Proprietors' Names.* 1879. 1 inch to 2 chains. MR 1/1810/2. The National Archives, Kew, UK.
———. *Plan of the Town of Lagos, West Coast of Africa: Prepared for Lagos Executive Commissioners of the Colonial and Indian Exhibition 1886; By Order of Fred. Evans, Esq. CMG. Deputy Governor.* 1885. About 1 inch to 275 feet (scale). CO 700/Lagos9/1. The National Archives, Kew, UK.
———. *Plan of the Town of Lagos, West Coast of Africa: Prepared for Lagos Executive Commissioners of the Colonial and Indian Exhibition 1886; By Order of Fred. Evans, Esq. CMG. Deputy Governor.* 1885. CO 700/Lagos9/2. The National Archives, Kew, UK.
Map of Isàlẹ̀ Èkó. In Isale Eko Descendants' Union, *Isale Eko Day, 2018.* Isale Eko Descendants' Union, 2018.
Map of the Colony of Lagos and Neighbouring Territories. 1888. 1 inch to 8 statute miles (scale). Compiled in the Intelligence Division, War Office, CO 700/ Lagos 15. The National Archives, Kew, UK.
Missing Admiralty Map. 1855. FO 84/976 Slave Trade No. 31. The National Archives, Kew, UK.
Outline Map Shewing the British Territory at Lagos (Now in Nigeria). July 22, 1871. MR 1/389/1. The National Archives, Kew, UK.
Peters, E. *Lagos Colony (Now Part of Nigeria): Ground and First Floor Plans of the Main Building.* May 28, 1879. MR 1/1766/12. The National Archives, Kew, UK.
———. *Lagos Colony (Now Part of Nigeria): Plan of Government House and Its Grounds, Outbuildings and Gardens.* May 28, 1879. MR 1/1766/11. The National Archives, Kew, UK.
Plan of Lagos. 1912 (Revised to 1926). Maps Y.484. The British Library, London.
Plan of Lagos. 1942. Maps X.1726. British Library, London.
Plan of Lagos and the Disputed Lands Ordered by Consul Campbell. 1855. Inclosure in Slave Trade No. 31. FO 84/976. The National Archives, Kew, UK.
Plan of Lagos Island, Lagos (Now in Nigeria), Showing the Layout of the Town, Lagoons, Reclaimed Land, and Iddo Island Connected to Lagos Island by Carter Bridge. 1907. 1 inch to 660 feet (scale). MPG 1/987/1. The National Archives, Kew, UK.
Plan of the Town of Lagos. Revised to May 1908. 1 inch to 5 chains (scale). CO 700/Lagos 31. The National Archives, Kew, UK.

Plan of the Town of Lagos. August 1911. 1 inch to 5 chains (330 feet; scale). MR Nigeria S.13. The Royal Geographical Society, London.
Plan of the Town of Lagos. 1913. 1 inch to 330 feet (scale). CO 1047/535/1. The National Archives, Kew, UK.
Plan of the Town of Lagos. 1917. CO 1047/535/1. The National Archives, Kew, UK.
Plan of the Town of Lagos (Revised to January 1926), Showing Principal Buildings, Roads, Tramways and District Boundaries. 1926. 1 inch to 330 feet (scale). CO 1047/677. The National Archives, Kew, UK.
Plan of the Town of Lagos—West Africa (in 15 Sheets). 1891. 1 inch to 88 feet (scale). CO 700/Lagos 14. The National Archives, Kew, UK.
Pratt, J. A., and E. P. Cotton. *Map of Lagos Town and Island (Now in Nigeria) with Circumjacent Features.* 1904. 4 inches to 1 mile (scale). MPG 1/987/2. The National Archives, Kew, UK.
Sketch Map of Abeokuta, Nigeria, Showing Its Position on the River Ogun, the Layout of the Town, Mission Stations, and Type of Country Surrounding the Settlement. July 1892. 2 inches to 1 mile. MPG 1/850/2. The National Archives, Kew, UK.
Sketch Map of Akitoye's Grant to the CMS (with a Sketch of the Compounds). 1855. TNA FO 84/976. The National Archives, Kew, UK.
Sketch Map of the Town of Lagos (Now in Nigeria): Plan Showing Original Swamp Districts, Present Swamps, and Sites of Camps. 1884. MPG 1/1162/9; one item extracted from CO 879/21 f.180. The National Archives, Kew, UK.
Sketch of Lagos and the Adjacent Country (To Accompany Report by Col. H. St. George Ord, R.E., 1865). 1865. CO 700/Lagos 1. The National Archives, Kew, UK.
Speeding, William C. *Lagos Harbour.* April 1898. 1 inch to 1,600 feet (scale), 63 × 90 cm. The National Archives, Kew, UK.
Town of Lagos: West Africa (15 Sheets Mounted as One). 1926. 1 inch to 88 feet (scale). CO 1047/676. The National Archives, Kew, UK.
The Traditional Oko Faji Layout. In Adams and Alli, *Oko Faji*, 28.
Ward, H. P. *A Plan of the River Lagos.* 1851. 3.5 inches to 1 sea mile (scale). FO 925/1052. The National Archives, Kew, UK.

Newspapers and Periodicals

The African Times (London)
The Anglo-African (Lagos)
The Church Missionary Gleaner (London)
The Church Missionary Intelligencer and Record (London)
The Illustrated London News
The Lagos Standard
The Lagos Weekly Record
The Spectator (Lagos)
The Times (London)
The Wesleyan-Methodist Magazine (London)

Adams, Alhaji Lawal Babatunde. *"Counting the Planets": The History and Development of OkePopo Community.* Lagos: Akuro Enterprises, 2003.

Adams, Lawal Babatunde. *The History, People and Culture of Ita-Tinubu Community.* Lagos: Tinubu Foundation, 2002.

Adams, John. *Sketches Taken during Ten Voyages to Africa between the Years 1786 and 1800, Including Observations on the Country between Cape Palmas and the River Congo, and Cursory Remarks on the Physical and Moral Character of the Inhabitants, with an Appendix Containing an Account of the European Trade with the West Coast of Africa.* London: Hurst, Robinson and Co.; Constable and Co. etc., 1822.

Adeleke, Tunde. *UnAfrican Americans: Nineteenth-Century Black Nationalists and the Civilizing Mission.* Lexington: University Press of Kentucky, 1998.

Adelusi-Adeluyi, Ademide. "Historical Tours of 'New' Lagos: Performance, Place Making, and Cartography in the 1880s." *Comparative Studies of South Asia, Africa and the Middle East* 38, no. 3 (2018): 443–54.

Aderinto, Saheed. *When Sex Threatened the State: Illicit Sexuality, Nationalism, and Politics in Colonial Nigeria, 1900–1958.* Champaign: University of Illinois Press, 2015.

Ajayi, J. F. Ade. *Christian Missions in Nigeria, 1841–1891: The Making of a New Élite.* Evanston, IL: Northwestern University Press, 1965.

———. "The British Occupation of Lagos, 1851–1861: A Critical Review." *Nigeria Magazine* 69 (1961): 96–105.

———. "Narrative of Samuel Ajayi Crowther." In *Africa Remembered: Narratives by West Africans from the Era of the Slave Trade,* edited by Philip D. Curtin, 289–316. Long Grove, IL: Waveland, 1997.

Akinsemoyin, Kunle, and Alan Vaughan-Richards. *Building Lagos.* 2nd ed. St. Helier, Jersey: Pengrail, 1977.

Akyeampong, Emmanuel Kwaku. *Between the Sea and the Lagoon: An Eco-social History of the Anlo of Southeastern Ghana, c. 1850 to Recent Times.* Athens: Ohio University Press, 2001.

Alli, Hon. Adekunle. "Lagos from the Earliest Times to British Occupation." Lecture presented at the Centre for Lagos Studies Distinguished Lecture Series, Adeniran Ogunsanya College of Education, Otto/Ijanikin, Lagos State, 2002.

Anderson, David M., and Richard Rathbone. "Introduction: Urban Africa; Histories in the Making." In *Africa's Urban Past,* edited by David M. Anderson and Richard Rathbone, 1–17. Oxford: James Currey, 2000.

Ayers, Edward L. "Turning toward Place, Space, and Time." In *The Spatial Humanities: GIS and the Future of Humanities Scholarship,* edited by David J. Bodenhamer, John Corrigan, and Trevor M. Harris, 1–13. Bloomington: Indiana University Press, 2010.

Baldwin, James. "Unnameable Objects and Unspeakable Crimes." In *The White Problem in America*, edited by the Editors of *Ebony*, 173–81. Chicago: Johnson Publishing, 1966.
Barber, Karin. *The Anthropology of Texts, Persons and Publics: Oral and Written Culture in Africa and Beyond*. Cambridge: Cambridge University Press, 2007.
Barrow, Ian J. *Making History, Drawing Territory: British Mapping in India, c. 1756–1905*. New Delhi: Oxford University Press, 2003.
Bascom, William. "Some Aspects of Yoruba Urbanism." *AMAN American Anthropologist* 64, no. 4 (1962): 699–709.
Belmessous, Saliha. *Native Claims: Indigenous Law against Empire, 1500–1920*. Oxford: Oxford University Press, 2012.
Bigon, Liora. *A History of Urban Planning in Two West African Colonial Capitals: Residential Segregation in British Lagos and French Dakar (1850–1930)*. London: Edwin Mellen Press, 2009.
———. "Urban Planning, Colonial Doctrines and Street Naming in French Dakar and British Lagos, c. 1850–1930." *Urban History* 36, no. 3 (2009): 426–48.
———. "Work in Progress: From French Senegal to British Nigeria; Fortier's Visit." *Historical Geography Research Group Newsletter*, Summer 2011.
Biobaku, Saburi O. *The Egba and Their Neighbours 1842–1872*. Oxford: Clarendon, 1957.
Blackett, R. J. M. "Martin R. Delany and Robert Campbell: Black Americans in Search of an African Colony." *Journal of Negro History* 62, no. 1 (1977): 1–25.
———. "Return to the Motherland: Robert Campbell, a Jamaican in Early Colonial Lagos." *Phylon* 40, no. 4 (1979): 375–86.
———. "Robert Campbell and the Triangle of the Black Experience." In *Beating against the Barriers: The Lives of Six Nineteenth-Century Afro-Americans*, 138–82. Baton Rouge: Louisiana University Press, 1986.
Bodenhamer, David J., John Corrigan, and Trevor M. Harris. *The Spatial Humanities: GIS and the Future of Humanities Scholarship*. Bloomington: Indiana University Press, 2010.
Bowen, T. J. *Central Africa: Adventures and Missionary Labors in Several Countries in the Interior of Africa, from 1849 to 1856*. Charleston, SC: Southern Baptist Publication Society, 1857.
Brantlinger, Patrick. "Freedom Burning: Anti-slavery and Empire in Victorian Britain." *Canadian Journal of History* 48 (2013): 339.
Brown, Spencer H. "Colonialism on the Cheap: A Tale of Two English Army Surgeons in Lagos, Samuel Rowe and Frank Simpson, 1862–1882." *International Journal of African Historical Studies* 27, no. 3 (1994): 551–88.
———. "A History of the People of Lagos, 1852–1886." PhD diss., Northwestern University, 1964.
Brown, Vincent. "Mapping a Slave Revolt: Visualizing Spatial History through the Archives of Slavery." *Social Text* 125 (December 2015): 134–41.

Brummett, Palmira. *Mapping the Ottomans: Sovereignty, Territory, and Identity in the Early Modern Mediterranean.* New York: Cambridge University Press, 2015.

Burke, Peter. *Eyewitnessing: The Uses of Images as Historical Evidence.* Picturing History Series. London: Reaktion Books, 2001.

Burns, Alan. *History of Nigeria.* London: Allen & Unwin, 1963.

Burton, Richard F. *Abeokuta and the Cameroons Mountains: An Exploration.* London: Tinsley Bros., 1863.

———. "A Day at Lagos." In *Wanderings in West Africa from Liverpool to Fernando Po: By a F. R. G. S. with Map and Illustration.* Vol. 2. London: Tinsley Brothers, 1863.

———. *A Visit to Gelele, King of Dahome: With Notices of the So Called "Amazons," the Grand Customs, the Yearly Customs, the Human Sacrifices, the Present State of the Slave Trade, and the Negro's Place in Nature.* Vol. 1. 2nd ed. London: Tinsley Brothers, 1864.

Callaway, Helen. "Spatial Domains and Women's Mobility in Yorubaland, Nigeria." In *Women and Space: Ground Rules and Social Maps,* edited by Shirley Ardener, 168–86. Providence: Berg Publishers, 1993.

Camp, Stephanie M. H. *Closer to Freedom: Enslaved Women and Everyday Resistance in the Plantation South.* Chapel Hill, NC: University of North Carolina Press, 2006.

Campbell, Robert. *A Few Facts, Relating to Lagos, Abbeokuta, and Other Sections of Central Africa.* Philadelphia: King & Baird, 1860.

———. *A Pilgrimage to My Motherland: An Account of a Journey among the Egbas and Yorubas of Central Africa, in 1859–60.* New York: Thomas Hamilton, 1861.

Candido, Mariana. *An African Slave Port and the Atlantic World: Benguela and Its Hinterland.* Cambridge: Cambridge University Press, 2015.

Cole, Patrick. "Lagos Society in the Nineteenth Century." In *Lagos: The Development of an African City,* edited by A. B. Aderibigbe, 27–58. Ikeja: Longman Nigeria, 1975.

———. *Modern and Traditional Elites in the Politics of Lagos.* Cambridge: Cambridge University Press, 1975.

Coleman, James Smoot. *Nigeria: Background to Nationalism.* Benin City: Broburg & Wiström, 1986.

Colonial and Indian Exhibition 1886. *Handbook to the West African Court: Gold Coast, Lagos, Sierra Leone, Gambia: With a Map; Compiled under the Direction of Sir James Marshall, C. M. G. Executive Commissioner.* London: William Cloves & Sons, 1886.

Craib, Raymond B. *Cartographic Mexico: A History of State Fixations and Fugitive Landscapes.* Durham, NC: Duke University Press, 2004.

Crowther, Samuel Ajayi. *Vocabulary of the Yoruba Language.* London: Printed for the Church Missionary Society, 1843.

———. *A Vocabulary of the Yoruba Language: Compiled by the Rev. Samuel Crowther, Native Missionary of the Church Missionary Society; Together with Introductory Remarks by the Rev. O. E. Vidal, M.A. Bishop Designate of Sierra Leone.* London: Seeleys, Fleet Street, 1852.

Cunha, Marianno Carneiro da, and Pierre Verger. *Da senzala ao sobrado: Arquitetura brasileira na Nigéria e na República Popular do Benim = From Slave Quarters to Town Houses: Brazilian Architecture in Nigeria and the People's Republic of Benin.* São Paulo: Nobel, 1985.

Darch, John H. *Missionary Imperialists? Missionaries, Government and the Growth of the British Empire in the Tropics 1860–1885.* Studies in Christian History and Thought. Milton Keynes, UK: Paternoster, 2009.

Delany, Martin Robinson. *Official Report of the Niger Valley Exploring Party.* New York: Webb, Millington; London: Thomas Hamilton, 1861.

Deniga, Adeoye. *Notes on Lagos Streets.* 3rd ed. Lagos: Jehovah Shalom / Tika-Tore Printing Press, 1921.

Dorn, Sherman. "Is (Digital) History More Than an Argument about the Past?" In *Writing History in the Digital Age,* edited by Jack Dougherty and Kristen Nawrotzki, 21–34. Ann Arbor: University of Michigan Press, 2013.

Echeruo, Michael J. C. *Victorian Lagos: Aspects of Nineteenth Century Lagos Life.* London: Macmillan, 1977.

Edelson, S. Max. *The New Map of Empire: How Britain Imagined America before Independence.* Cambridge, MA: Harvard University Press, 2017.

Edney, Matthew H. *Mapping an Empire: The Geographical Construction of British India, 1765–1843.* Chicago: University of Chicago Press, 1997.

Elebute, Adeyemo. *The Life of James Pinson Labulo Davies: A Colossus of Victorian Lagos / Adeyemo Elebute.* 2nd ed. Lagos: Prestige, 2017.

Falola, Toyin, and Dare Oguntomisin. "Prince Kosoko." In *Yoruba Warlords of the Nineteenth Century,* 141–58. Trenton, NJ: Africa World Press, 2001.

Fasinro, Alhaji H. A. B. *Political and Cultural Perspectives of Lagos.* Lagos: Academy Press PLC, 2004.

Ferreira, Roquinaldo Amaral. *Cross-Cultural Exchange in the Atlantic World: Angola and Brazil during the Era of the Slave Trade.* New York: Cambridge University Press, 2012.

Foote, Mrs. Henry Grant. *Recollections of Central America and the West Coast of Africa: By Mrs. Foote, Widow of the Late Henry Grant Foote.* London: T. Cautley Newby, 1869.

Forbes, Frederick E. *Dahomey and the Dahomans: Being the Journals of Two Missions to the King of Dahomey, and Residence at His Capital, in the Years 1849 and 1850.* 2 vols. London: Cass, 1966.

Fourchard, Laurent. "Between World History and State Formation: New Perspectives on Africa's Cities." *Journal of African History* 52, no. 2 (2011): 223–48.

Fraser, Louis. *Dahomey and the Ending of the Trans-Atlantic Slave Trade: The Journals and Correspondence of Vice-Consul Louis Fraser, 1851–1852.* Edited by Robin Law. Oxford: Oxford University Press, 2012.

Freedom Park: The Journey of Liberation; A Commemorative Publication to Mark the Commissioning of Freedom Park, the Transformed Old Broad

Street Prison as Part of Lagos State's Official Celebration of Nigeria at 50. Lagos: M2DC for Hermitage Publishing, 2010.
"Fundamentals of Georeferencing a Raster Dataset." ArcGIS for Desktop. ESRI. Accessed March 7, 2019. https://desktop.arcgis.com/en/arcmap/10.3/manage-data/raster-and-images/fundamentals-for-georeferencing-a-raster-dataset.htm.
Gale, Thomas S. "Lagos: The History of British Colonial Neglect of Traditional African Cities." *African Urban Studies* 5 (Fall 1979): 11–24.
Gbadamosi, T. G. O. *The Growth of Islam among the Yoruba, 1841–1908.* Ibadan History Series. Atlantic Highlands, NJ: Humanities Press, 1978.
Geary, Sir William N. M. *Nigeria under British Rule.* New York: Barnes & Noble, 1965.
George, Abosede A. *Making Modern Girls: A History of Girlhood, Labor, and Social Development in Colonial Lagos.* Athens: Ohio University Press, 2014.
Glaeser, Edward L. "What Can Developing Cities Today Learn from the Urban Past?" Working Paper Series, National Bureau of Economic Research, 2021.
Glover, Lady Elizabeth Rosetta. *Life of Sir John Hawley Glover, R.N. G.C.M.G.* London: Smith, Elder, 1897.
Godwin, John, and Gillian Hopwood. *Sandbank City: Lagos at 150.* Lagos: Prestige, 2012.
Gollmer, Charles, Jr. *Charles Andrew Gollmer: His Life and Missionary Labours in West Africa: Compiled from His Journals and the CMS's Publications.* London: Hodder and Stoughton, 1889.
———. "Rebellion at Lagos—Deposition of the King (Western Africa, C.M.S.)." *Missionary Register* 34 (1846): 436–46.
———. "Western Africa, Extracts from the Journal of Rev. Charles Gollmer." *Missionary Register* 34, no. 6 (June 1846).
Gregory, Ian N., and Alistair Geddes, eds. *Toward Spatial Humanities: Historical GIS and Spatial History.* Bloomington: Indiana University Press, 2014.
Harley, J. B. "Deconstructing the Map." In *Classics in Cartography: Reflections on Influential Articles from Cartographica,* edited by Martin Dodge, chapter 16. Chichester, UK: J. Wiley & Sons, 2011.
Healy, John J. C. "Land Tenure in the Colony of Lagos." In *Land Tenure in West Africa: Reports by T. C. Rayner, Esq., Chief Justice of Lagos, and J. J. C. Healy, Esq., Land Commissioner,* 1–14. Foreign and Commonwealth Office Collection, 1897.
Holdich, T. H. "How Are We to Get Maps of Africa?" *Geographical Journal* 38, no. 6 (December 1901): 590–601.
Hopkins, Antony G. "Property Rights and Empire Building: Britain's Annexation of Lagos, 1861." *Journal of Economic History* 40, no. 4 (December 1980): 777–98.
Howard, Allen M., and Richard M. Shain, eds. *The Spatial Factor in African History: The Relationship of the Social, Material, and Perceptual.* Boston: Brill, 2005.

Huzzey, Richard. *Freedom Burning: Anti-slavery and Empire in Victorian Britain.* Ithaca, NY: Cornell University Press, 2012.

Ikime, Obaro. *The Fall of Nigeria: The British Conquest.* London: Heinemann, 1977.

Irving, Robert Forsyth. *A Collection of the Principal Enactments and Cases Relating to Titles to Land in Nigeria, Compiled in Connection with the Sale of Enemy Properties, Revised, with Introduction and Notes.* London: Stevens and Sons, 1916.

Isale Eko Descendants Union. *Isale Eko Day, 2018.* Isale Eko Descendants Union, 2018.

Jacobs, Allan B. *Great Streets.* Cambridge, Mass.: The MIT Press, 1993.

Johnson, Jessica Marie. *Wicked Flesh: Black Women, Intimacy, and Freedom in the Atlantic World.* Philadelphia: University of Pennsylvania Press, 2020.

Johnson, Rashauna. *Slavery's Metropolis: Unfree Labor in New Orleans during the Age of Revolutions.* Cambridge, MA: Cambridge University Press, 2016.

Jones, Adam. *German Sources for West African History, 1599–1669.* Studien Zur Kulturkunde. Wiesbaden: Franz Steiner Verlag GmbH, 1983.

Knowles, Anne Kelly, and Amy Hillier, eds. *Placing History: How Maps, Spatial Data, and GIS Are Changing Historical Scholarship.* Redlands, CA: ESRI, 2008.

Kopytoff, Jean Herskovits. *A Preface to Modern Nigeria: The "Sierra Leonians" in Yoruba, 1830–1890.* Madison: University of Wisconsin Press, 1965.

Kọtun, Prince Bọlakalẹ. *A History of the Eko Dynasty.* Abuja: National Assembly Printing Press, 1998.

Law, Robin. "Between the Sea and the Lagoons: The Interaction of Maritime and Inland Navigation on the Precolonial Slave Coast." *Cahiers d'Études Africaines* 29 (1989): 209–37.

———. "The Career of Adele at Lagos and Badagry, c. 1807–c. 1837." *Journal of the Historical Society of Nigeria* 9, no. 2 (1978): 35–59.

———, ed. *From Slave Trade to "Legitimate" Commerce: The Commercial Transition in Nineteenth-Century West Africa.* Cambridge, MA: Cambridge University Press, 2007.

———. "On Pawnship and Enslavement for Debt in the Pre-colonial Slave Coast." In *Pawnship in Africa: Debt Bondage in Historical Perspective,* edited by Paul Lovejoy and Toyin Falola, 51–69. Boulder, CO: Westview, 1994.

———. "Ouidah: A Pre-colonial Urban Centre in Coastal West Africa, 1727–1892." In *Africa's Urban Past,* edited by David M. Anderson and Richard Rathbone, 85–97. Oxford: James Currey, 2000.

———. *Ouidah: The Social History of a West African Slaving "Port," 1727–1892.* Athens: Ohio University Press, 2004.

———. "Trade and Politics behind the Slave Coast: The Lagoon Traffic and the Rise of Lagos, 1500–1800." *Journal of African History* 24 (1983): 321–48.

Lawal, B. Adams, and Hon. Adekunle Alli, eds. *Oko Faji: The Biography of a Community.* Lagos: CreateSpace Independent Publishing Platform, 2017.

Lawrance, Benjamin N., Emily Lynn Osborn, and Richard L. Roberts. "Introduction: African Intermediaries and the 'Bargain' of Collaboration." In

Intermediaries, Interpreters, and Clerks: African Employees in the Making of Colonial Africa, edited by Benjamin N. Lawrance, Emily Lynn Osborn, and Richard L. Roberts, 3–34. Africa and the Diaspora. Madison: University of Wisconsin Press, 2006.

Lefebvre, Henri. *The Production of Space.* Oxford: Blackwell, 1991.

Lindsay, Lisa A. *Atlantic Bonds: A Nineteenth-Century Odyssey from America to Africa.* H. Eugene and Lillian Youngs Lehman Series. Chapel Hill: University of North Carolina Press, 2017.

———. "'To Return to the Bosom of Their Fatherland': Brazilian Immigrants in Nineteenth-Century Lagos." *Slavery & Abolition: A Journal of Slave and Post-slave Studies* 15, no. 1 (1994): 22–50.

Lọṣi, John B. Prince. *History of Lagos.* 1914. Reprint, Lagos: African Education Press, 1967.

Lọṣi, Ọmọ-Ọba John B. *Itan Eko.* Lagos: Church Missionary Society Bookshop, 1934.

Lovejoy, Paul. "Pawnship and Seizure for Debt in the Process of Enslavement in West Africa." In *Debt and Slavery in the Mediterranean and Atlantic Worlds,* edited by Gwyn Campbell and Alessandro Stanziani, 63–75. Financial History 22. London: Pickering & Chatto, 2013.

Lucas, Archdeacon J. Olumide. *St. Paul's Church, Breadfruit Lagos: Lecture on the History of the Church (1852–1945).* N.p.: 1946.

Mabogunje, Akin L. "Lagos, a Study in Urban Geography." PhD diss., University of London, 1961.

———. "Lagos—Nigeria's Melting Pot." *Nigeria Magazine,* August 1961, 128–55.

———. "Overview of Research Priorities in Africa." In *Urban Research in the Developing World,* edited by Richard Stren, 21–45. Toronto: University of Toronto, 1994.

———. *Urbanization in Nigeria.* New York: Africana Pub., 1969.

Macaulay, Herbert Samuel Heclas. *Justitia Fiat: The Moral Obligation of the British Government to the House of King Docemo of Lagos; An Open Letter.* London: Printed by St. Clements Press, 1921.

———. *The Lagos Land Question: Speech Delivered on the Occasion by Mr. Herbert Macaulay, at Government House, Lagos, 13th June, 1912.* Lagos, 1912.

Macaulay, Zachary. "Cuban Slaves in England: Testimony of Ignatio Moni." *The Anti-slavery Reporter: Under the Sanction of the British and Foreign Anti-slavery Society* (October 1854): 234–39.

Mann, Kristin. "African and European Initiatives in the Transformation of Land Tenure in Colonial Lagos (West Africa), 1840–1920." In *Native Claims: Indigenous Law against Empire, 1500–1920,* 223–47. Oxford: Oxford University Press, 2012.

———. "Interpreting Cases, Disentangling Disputes: Court Cases as a Source for Understanding Patron-Client Relations in Early Colonial Lagos." In *Sources and Methods in African History: Spoken, Written, Unearthed,* edited by Toyin Falola and Christian Jennings, 195–218. Rochester, NY: University of Rochester Press; Woodbridge, UK: Boydell & Brewer, 2003.

———. *Marrying Well: Marriage, Status, and Social Change among the Educated Elite in Colonial Lagos.* African Studies Series 47. Cambridge: Cambridge University Press, 1985.

———. "The Rise of Taiwo Olowo: Law, Accumulation and Mobility in Early Colonial Lagos." In *Law in Colonial Africa,* edited by Kristin Mann and Richard L. Roberts, 85–102. Social History of Africa. Portsmouth, NH: Heinemann Educational Books, 1991.

———. *Slavery and the Birth of an African City: Lagos, 1760–1900.* Bloomington: Indiana University Press, 2007.

Mann, Kristin, and Richard L. Roberts, eds. *Law in Colonial Africa.* Social History of Africa. Portsmouth, NH: Heinemann Educational Books, 1991.

Manning, Patrick. "Coastal Society in the Republic of Bénin: Reproduction of a Regional System (La Société Côtière de La République Du Bénin: Reproduction d'un Système Régional)." *Cahiers d'Études Africaines* 29, no. 114 (1989): 239–57.

Mares, Detlev, and Wolfgang Moschek. "Place in Time: GIS and the Spatial Imagination in Teaching History." In *History and GIS,* edited by Alexander von Lünen and Charles Travis, 59–72. Dordrecht, the Netherlands: Springer, 2013.

Martínez-Fernández, Luis. *Fighting Slavery in the Caribbean: The Life and Times of a British Family in Nineteenth-Century Havana.* Armonk, NY: M. E. Sharpe, 1998.

Matheson, Jane Diane. "Lagoon Relations in the Era of Kosoko, 1845–1862: A Study of African Reaction to European Intervention." PhD diss., Boston University, 1974.

Mba, Nina. "Women in Lagos Political History." In *History of the Peoples of Lagos State,* edited by Ade Adefuye, Babatunde Agiri, and Jide Osuntokun, 233–45. Lagos: Lantern Books, 2002.

Mbembe, Achille. *On the Postcolony.* Berkeley: University of California Press, 2001.

Mbembe, Achille, and Sarah Nuttall. "Writing the World from an African Metropolis." *Public Culture Public Culture* 16, no. 3 (2004): 347–72.

McIntyre, W. D. "Commander Glover and the Colony of Lagos, 1861–73." *Journal of African History* 4, no. 1 (March 1963): 57–79.

Miers, Suzanne, and Igor Kopytoff. *Slavery in Africa: Historical and Anthropological Perspectives.* Madison: University of Wisconsin Press, 1977.

Mitchel, N. C. "Yoruba Towns." In *Essays on African Population,* edited by K. M. Barbour and R. Mansell Prothero, 279–301. London: Routledge and Paul, 1961.

M'Leod, John. *A Voyage to Africa with Some Account of the Manners and Customs of the Dahomian People.* London: Cass, 1971.

Monmonier, Mark S. *Drawing the Line: Tales of Maps and Cartocontroversy.* New York: H. Holt, 1995.

———. *How to Lie with Maps.* Chicago: University of Chicago Press, 1991.

Morton, David. *Age of Concrete: Housing and the Shape of Aspiration in the Capital of Mozambique.* Athens: Ohio University Press, 2019.

Mundy, Barbara E. *The Mapping of New Spain: Indigenous Cartography and the Maps of the Relaciones Geográficas.* Chicago: University of Chicago Press, 1996.

Nead, Lynda. *Victorian Babylon: People, Streets and Images in Nineteenth-Century London.* New Haven, CT: Yale University Press, 2000.

Nigeria Law Reports. "Sakariyawo Oshodi v. Moriamo Dakolo and Others." In *A Selection of Cases Decided in Divisional Courts of Nigeria during the Years 1928 and 1929, Together with the Privy Council and Full Courts Appeals Relating to Cases Falling within These Two Years.* Vol. 9. Lagos: Government Printer, 1932.

Nwanunobi, C. Onyeka. "Incendiarism and Other Fires in Nineteenth-Century Lagos (1863–88)." *Africa: Journal of the International African Institute* 60 (1990): 111–20.

O'Dowd, Mary, and June Purvis, eds. *A History of the Girl.* New York: Springer International Publishing, 2018.

Ojo, Olatunji. "The Business of 'Trust' and the Enslavement of Yoruba Women and Children for Debt." In *Debt and Slavery in the Mediterranean and Atlantic Worlds,* edited by Gwyn Campbell and Alessandro Stanziani, 77–91. Financial History 22. London: Pickering & Chatto, 2013.

———. "Document 2: Letters Found in the House of Kosoko, King of Lagos (1851)." *African Economic History* 40, no. 1 (2012): 37–126.

———. "The Yoruba Church Missionary Society Slavery Conference 1880." *African Economic History* 49, no. 1 (2021): 73–103.

Olukoju, Ayodeji. "The Cost of Living in Lagos, 1914–45." In *Africa's Urban Past,* edited by David M. Anderson and Richard Rathbone, 126–43. Oxford: James Currey, 2000.

———. *Infrastructure Development and Urban Facilities in Lagos, 1861–2000.* Ibadan, Nigeria: Institut Français de Recherché en Afrique, University of Ibadan, 2003.

———. *The Liverpool of West Africa: The Dynamics and Impact of Maritime Trade in Lagos, 1900–1950.* Trenton, NJ: Africa World, 2005.

———. "Population Pressure, Housing and Sanitation in West Africa's Premier Port City: Lagos, 1900–1939." *Great Circle* 15, no. 2 (1993): 91–106.

———. "Which Lagos, Whose (Hi)Story?" *Lagos Notes and Records* 24, no. 1 (2018): 140–70.

Olukoju, Ayodeji, Saheed Aderinto, and Paul Osifodunrin. *The Third Wave of Historical Scholarship on Nigeria Essays in Honor of Ayodeji Olukoju.* Newcastle, UK: Cambridge Scholars, 2012.

Omu, Fred I. A. "The Anglo-African 1863–65." *Nigeria Magazine,* September 1966, 206–12.

———. *Press and Politics in Nigeria, 1880–1937.* Atlantic Highlands, NJ: Humanities Press, 1978.

Osifodunrin, Paul. *The First Church in Lagos: History of Holy Trinity Church, Ebute Ero, 1852–2016.* Ibadan: University Press PLC, 2018.

Pamuk, Orhan. *Istanbul: Memories of a City.* Translated by Maureen Freeley. London: Faber and Faber, 2005.

Payne, John Augustus Otonba. *Payne's Lagos and West African Almanack and Diary.* London: J. S. Phillips, 1882.

———. *Payne's Lagos and West African Almanack and Diary.* London: J. S. Phillips, 1883.

———. *Payne's Lagos and West African Almanack and Diary.* London: J. S. Phillips, 1894.

———. *Table of Principal Events in Yoruba History, with Certain Other Matters of General Interest, Compiled Principally for Use in the Courts within the British Colony of Lagos, West Africa.* Lagos: Printed by Andrew M. Thomas, 1893.

Quayson, Ato. *Oxford Street: City Life and the Itineraries of Transnationalism.* Durham, NC: Duke University Press, 2014.

Rankin, William. *After the Map Cartography, Navigation, and the Transformation of Territory in the Twentieth Century.* Chicago: University of Chicago Press, 2018.

Rayfield, J. R. "Theories of Urbanization and the Colonial City in West Africa." *Africa* 44, no. 2 (April 1974): 163–85.

Rayner, T. C. "Land Tenure in West Africa." In *Land Tenure in West Africa: Reports by T. C. Rayner, Esq., Chief Justice of Lagos, and J. J. C. Healy, Esq., Land Commissioner,* 1–5. Foreign and Commonwealth Office Collection, 1897.

Ross, David A. "The Career of Domingo Martinez in the Bight of Benin 1833–64." *Journal of African History* 6 (1965): 79–90.

Roy, Ananya. "Why India Cannot Plan Its Cities: Informality, Insurgence and the Idiom of Urbanization." *Planning Theory* 8, no. 1 (February 1, 2009): 76–87.

Shasore, Olasupo. *Possessed: A History of Law and Justice in the Crown Colony of Lagos 1861–1906.* Nigeria: Quramo Books, 2014.

Simone, Abdul Malik. *For the City Yet to Come: Changing African Life in Four Cities.* Durham, NC: Duke University Press, 2006.

———. "People as Infrastructure." In *Johannesburg: The Elusive Metropolis,* edited by Achille Mbembe and Sarah Nuttall, 68–90. Durham, NC: Duke University Press, 2008.

Smith, Robert Sydney. *The Lagos Consulate: 1851–1861.* Berkeley: University of California Press, 1979.

Solnit, Rebecca, and Rebecca Snedeker. *Unfathomable City: A New Orleans Atlas.* Berkeley: University of California Press, 2013.

Sorenson-Gilmour, Caroline. "Slave Trading along the Lagoons of South West Nigeria: The Case of Badagry." In *Ports of the Slave Trade (Bights of Benin and Biafra): Papers from a Conference of the Centre of Commonwealth Studies, University of Stirling, June 1998,* edited by Robin Law and Silke Strickrodt, 84–95. Stirling, UK: University of Stirling, 1999.

Stoler, Ann Laura, ed. *Imperial Debris: On Ruins and Ruination.* Durham, NC: Duke University Press, 2013.

Tamuno, Tekena N. *The Police in Modern Nigeria, 1861–1965: Origins, Development and Role.* Ibadan: Ibadan University Press, 1970.

Tew, Sir Meryn L. *Report to Title to Land in Lagos: Together with a Report of a Committee Set Up to Advise the Governor in Regard Thereto, and Draft Legislation to Give Effect to Certain Recommendations Contained Therein.* Lagos: Government Printer, 1939.

Tickell, Alex. "Negotiating the Landscape: Travel, Transaction, and the Mapping of Colonial India." *Yearbook of English Studies* 34 (2004): 18–30.

Townsend, George. *Memoir of the Rev. Henry Townsend, Late C.M.S. Missionary, Abeokuta, West Africa.* London: Marshall Bros.; Exeter, UK: J. Townsend, 1887.

Verger, Pierre. *Trade Relations between the Bight of Benin and Bahia from the 17th to 19th Century.* Ibadan: Ibadan University Press, 1976.

Wainaina, Binyavanga. "Wangechi Mutu Wonders Why Butterfly Wings Leave Powder on the Fingers, There Was a Coup Today in Kenya." *Jalada Africa* (blog), October 17, 2014.

Watson, Ruth. *"Civil Disorder Is the Disease of Ibadan": Chieftaincy and Civic Culture in a Yoruba City.* Western African Studies. Athens: Ohio University Press; Oxford: J. Curry; Ibadan: Heinemann Educational Books (Nigeria), 2003.

Whiteman, Kaye. *Lagos: A Cultural History.* Northampton, MA: Interlink Books, 2014.

Whitford, John. *Trading Life in Western and Central Africa.* Liverpool: "Porcupine" Office, Cable Street, 1877.

Wood, Rev. J. B. *Historical Notices of Lagos, West Africa and on the Inhabitants of Lagos: Their Character, Pursuits, and Languages.* Lagos: Church Missionary Society Bookshop, 1933.

Yemitan, Oladipo. *Madame Tinubu: Merchant and King-Maker.* Ibadan: University Press, 1987.

Index

The letter *f* following a page number denotes a figure; the letter *m*, a map.